DARWIN'S
APOSTLES:

THE MEN WHO FOUGHT TO HAVE EVOLUTION ACCEPTED, THEIR TIMES, AND **HOW THE BATTLE CONTINUES**

DARWIN'S
APOSTLES:

THE MEN WHO FOUGHT TO HAVE EVOLUTION ACCEPTED, THEIR TIMES, AND **HOW THE BATTLE CONTINUES**

David I. Orenstein, BA, MLS, MS, PhD
and Abby Hafer, BA, DPhil

Humanist
PRESS

HUMANISTPRESS.COM • WASHINGTON, DC

© 2019 Humanist Press LLC
1821 Jefferson Place NW
Washington, DC 20036
www.humanistpress.com

David Orenstein and Abby Hafer
Darwin's Apostles: The Men Who Fought to Have Evolution Accepted,
Their Times, and How the Battle Continues

Published by: Humanist Press LLC
Editor: Fred Edwords
Cover and interior design: Sharon McGill
Print ISBN: 978-0-931779-82-4
Ebook ISBN: 978-0-931779-83-1

The authors wish to dedicate this book to the researchers and teachers on the front lines of scientific investigation and science education. Without your hard work, dedication, and expertise our world and our lives would surely be darkened by ignorance, folly, and the fear of the unknown.

CONTENTS

ACKNOWLEDGEMENTS

From David I. Orenstein

Gestation is a fascinating topic for any biologist or field anthropologist. How long it takes for an animal to maturate; for a being, language, or society to evolve; or for a book on Charles Darwin to be written each have their own distinct lines of both maturation and gestation. In the case of *Darwin's Apostles*, the gestation was more than two years of interviews, research, writing, and editing to finally reach publication. That's about the same time it takes for an African elephant or sperm whale to be born. Or put into a different gestational context, that could be twenty-three chipmunks, two mountain gorillas, or thirty-two laboratory mice.

Our human time isn't based typically on gestation but the calendar year. And over the more than two years there have been many people who have given their time, energy, and support to this book project. Firstly, I wish to thank my coauthor, Abby Hafer, for her friendship and for agreeing to join me in this exploration of Darwin and his advocates. When I asked her to be my coauthor she was ferocious in her agreement that a book such as this needed to be written.

I wish to also thank three important people from "across the pond" for helping me with my research while in the United Kingdom. My friends, Sue and Gez Cox, allowed me to use their home as my "thinking office" while doing original research in 2016. It was my good fortune to contact Virginia Mills, the project coordinator of the Hooker Correspondence Project, for connecting me to the Hooker family. My deepest and most sincere thanks to Isobol Moses, the great-great-granddaughter of Joseph Dalton Hooker, for offering herself to be interviewed and for exploring what an honor it is to be a family relation—even a hundred years out—to such a great man of science.

A note of thanks to my colleagues and the administration at Medgar Evers College (CUNY) for their support of my research and for offering me such a wonderful place to teach Darwin and anthropology. MEC is a dream come true for me. A big thank you to Luis Granados, the senior editor at Humanist Press, for supporting the project from its inception. And a huge thanks to Fred Edwords and the Humanist Press staff for taking the project over the goal line. A fond note of thanks to my colleagues Dr. Owen Brown, Dr. Maria DeLongoria, and Dr. Ethan Gologor for their support of my teaching, activism, and work within the City University of New York. I would also like to thank Nick Fish, president of American Atheists, Inc., for allowing me to use their exceptional library and archives as my "man cave" to complete the book.

Finally, it would be impossible not to give thanks to my family: those who live with and put up with me daily, those across the United States, and the artists and great thinkers in Romania. Each group of individuals have supported and enlightened me in ways too numerous to put into a paragraph. But without their love and guidance I would not be the happy and healthy human being that I am today.

From Abby Hafer

The writing of this book has required patience, encouragement, and good cheer from those around me. My husband Alan MacRobert has supplied these in large and loving quantities. My coauthor David Orenstein has been unceasing in his encouragement and support. My employer Curry College gave me a one-course release in order to provide me with more time to work on this book.

Writing this book has not been easy but it has been interesting. With this in mind I want to thank the apostles themselves, even though they are not around to enjoy my praise. Thomas Henry Huxley, Joseph Dalton Hooker, Asa Gray, John William Draper, and Alfred Russel Wallace have made fine and entertaining companions. But I would also like to thank the apostles' apostles—those numerous entities that keep alive and offer to the public the words and deeds of these fascinating men and those around them.

The Darwin Correspondence Project (https://www.darwinproject.ac.uk/) has made available vast quantities of Charles Darwin's correspondence, in searchable form, by date, and by correspondent, with comments. That particular circle of scientists communicated a lot by post, so reading their letters seems almost like listening in on their conversations. There are similar organizations for some of the other apostles. The Alfred Russel Wallace Correspondence Project, founded by Dr. George Beccaloni (http://wallaceletters.info/content/homepage) does much the same thing for Wallace. And the Huxley File, (https://mathcs.clarku.edu/huxley/) created by Charles Blinderman and David Joyce at Clark University in Worcester, Massachusetts, is a trove of precious information about Thomas Henry Huxley. Kew Gardens has numerous different parts to its website that provided me with information not only about Joseph Hooker himself but also about military and economic aspects of botany and empire. The American Chemical Society provided me with useful information about John William Draper. And finally, the Arnold Arboretum—not as important internationally as Kew but beloved by Bostonians—provided me with wonderful information about Asa Gray himself, and about the Asa Gray Disjunction, about which research is still being done. I thank them for continuing to be obsessed with this fascinating topic.

In addition, if you want to find out more about evolution, then you want to visit the website of the University of California (at Berkeley's) Museum of Paleontology (UCMP). There, you can find information about scientists involved in evolutionary study, including but not limited to the apostles, and you can also visit UCMP's site *Understanding Evolution* (https://evolution.berkeley.edu/evolibrary/home.php), which is a great free online guide to the science of evolution itself. I thank the University of California for building and maintaining this invaluable public service.

To all of these far-flung idealistic providers of information—thank you! I couldn't have done it without you!

AUTHOR'S NOTES

No one on the planet can call me a religious person. I am an avowed atheist and humanist. In 2018, I began what will be an annual trip to the Amazon rainforest and visits to the great Galápagos Islands in Ecuador. I can write from experience that such natural beauty and biodiversity is breathtaking and deeply inspiring. The boat trips between the Galápagos Islands, the hikes through the Amazon rainforest, meeting all the wonderful people; the Ecuadorian guides, the many indigenous people who opened their homes and tables to me, meeting fellow travelers and working with my students, have each been a spiritual experience much like having religious faith.

Like Albert Einstein, Baruch Spinoza, David Hume, and Carl Sagan, my religion is nature and it requires no god and only the mechanics of the cosmos to be true. My belief in the human spirit has been encouraged and reaffirmed by my experiences in Ecuador. Such trips remind me that we are a small and fragile species and we owe it to ourselves, to future generations, and our evolution to be kind to one another.

Since the inception of this book I have labored to explore Charles Darwin's life. I have visited his home in Kent, his resting place inside Westminster Abbey, and have read so much by and about him to garner only deeper respect for his life, his loves, and his contributions to science and education.

Since Darwin's cause was taken up by so many brave companions, I have also sought out their lives as well, visiting and paying my respects to their graves. (Gray is in Boston; Wallace, Hooker, and Huxley in England, and Draper is buried not more than five miles from me in Park Slope, Brooklyn, New York.) I have read their work and followed their ideas and experiences in the Amazon rainforest and the laboratories at the Royal Kew Gardens. I have studied their adventures in the halls of academia and as they traveled up the sides of mountains; as they labored in the gardens and plains of the United States, South America, and England; and as they spoke out, wrote, and debated to secure our very modernity.

To think that Darwin spent just five weeks of his five year journey in the Galápagos Islands is itself stunning. He spent only nineteen days on land between four of the islands. Yet, this 1 percent of his time solidified his views on natural selection. That is the raw power of this place. The Galapagos exposes us to the reality of how the earth has operated since life first appeared some 3.5 billion years ago.

Darwin unlocked nature's secret for the wealth of human knowledge and human good. He realized this early on, while his supporters then and now acknowledge his contributions to our modernity. Darwin's detractors, forever a fearful lot, cannot stop working to unseat his grand laws for the sake of their own ill-formed comfort and uninformed philosophies. But it has been my privilege and honor to have spent a small part of my own life walking in Darwin's footsteps. It is my delight to teach his ideas. It remains my duty to advocate for science and science education, and to see the world as it is rather than how I'd hope it would be in my own lifetime.

This is my reality. It is how my conscious brain interprets the universe. It is my emotional connection to all that I know, to the people I encounter and attempt to understand, and to love.

And I owe this all to anthropology and to life itself.

David Orenstein
Professor of Anthropology
Medgar Evers College of the
City University of New York
July 25, 2018

AUTHOR'S NOTES

"I happen to have in my pocket a monkey's brain."

How can you not love a story with that line in it? A true story at that!

When David Orenstein asked me if I wanted to join him in this project, I sought to find out what I was getting myself into before I made a decision. I looked at the names of the apostles and started doing research on one I didn't know anything about—the botanist Asa Gray. I looked up materials online and discovered a fascinating article about the Asa Gray Disjunction in *Wired* magazine. So fascinated was I that I found myself not just eating my lunch in front of the computer so that I could keep reading, but actually brushing and flossing my teeth there as well! I was hooked, and I had to admit it. Perhaps only a biologist could get that excited by a story about international plant distributions, but I am one, and I was. The fact that Darwin managed to use Gray's own data to convince the reluctant botanist about the reality of evolution by natural selection made the story even better. The fact that a genuine heretofore unsolved and tenacious scientific mystery wound up being solved because Gray accepted evolution—well, what more could a girl ask for?

Then in reading about Huxley I found that his arguments in favor of evolution included fossils, which everybody loves, or if they don't they should; *and* classification, about which many biologists care deeply even though it is a mystery to most other people; *and* anatomical dissection, something which I have spent much of my career doing. I frequently say there is nothing like a good dissection to brighten your day, and clearly Huxley was a kindred spirit. Huxley did all these things and more. He engaged in the Victorian equivalent of a Twitter war, used plenty of snark, and managed to throw shade on a bishop. What's not to love?

And the fulcrum of his Victorian slanging match was that he accused his nemesis of doing (or at least carelessly using data from) botched dissections! Who knew that monkey brains could be so important? He even got friends to help! The man with the conveniently-placed monkey's brain in his pocket was not Huxley himself but a confederate named William Henry Flower. Said monkey's brain was brought to a public presentation given by someone else, who was saying a lot of wrong things about monkeys' brains in that very presentation. The monkey brain was then deployed and given a public dissection, thereby proving the mendacious presenter to have been wrong, and wrong for a very long time. And all this was in the aid of getting evolution accepted. Scientific meetings are rarely exciting in quite this way, but this is the stuff of legend.

How could I resist? A broadly-educated biologist like me who cares about plant distribution *and* evolution *and* cutting things up rarely gets to exercise quite as many of her faculties in one project.

The other apostles are equally charming. There's Draper, who was fascinated by chemistry and light, and came close to inventing the electric light bulb well before Edison, but he was too busy doing other things; and Hooker, who also saw and made full use of the connection between evolution and plant distributions and was Darwin's bestie as well; and Wallace, who has a special place in my heart for figuring out evolution using *animal* distributions (as Darwin seems to have done) *and* figuring out the ancient geology of an entire region based on those animal distributions. I love it when biology actually tells *other* fields of science what they need to know.

So I hope that you have as much fun reading about these apostles as I have had researching and writing about them.

Abby Hafer
Senior Lecturer in Biology
Curry College
Milton, MA
November 19, 2018

CHAPTER 1

THE DEBATE THAT USHERED IN MODERNITY

"There is no great invention, from fire to flying, which has not been hailed as an insult to some god."[1]

J. B. S. Haldane (1892–1964)
Mathematician and Evolutionary Biologist

The Legendary Debate

The scene is famous. It is June 30, 1860, at Oxford University. The British Association for the Advancement of Science had been having its annual meeting for several days. Charles Darwin's ground-breaking book, *On the Origin of Species* (henceforth referred to as simply *Origin*), had been published seven months earlier. The infamous Oxford evolution debate was taking place![2]

The room was packed and hot. Somebody fainted and had to be carried out. Legend has it that the fainting took place after Thomas Henry Huxley threw shade on Bishop Samuel Wilberforce of Oxford, but no one knows if that was actually the reason.[3] What we do know is that it was hot, packed, and tense. We also know that the *idea* that someone would needle a bishop in public was shocking in that time and place.

The atmosphere was not like a scientific meeting today—more like a cross between a Pittsburgh Steelers football game and a charismatic church service. This included having a group of women packed together, ready to wave white handkerchiefs in victory.[4] Anti-evolutionists included uniformed clergy in white clerical collars and black coats who were seated en masse right in the middle of the room.[5] They laughed and cheered loudly. This Victorian version of an amen corner was conspicuous in its numbers and decibel level, and provided seemingly genial cover for intimidation. A small group of undergraduates who supported evolution were shoved off to one corner of the great room. People had been choosing sides between the bishops and the biologists for months.[6]

It was understood that moral ascendancy was paramount, and that emotion could easily carry the day, especially since the anti-Darwinists were happy to pack the room with amateur enthusiasts whose main job was to be rowdy. The meeting would be raucous and loud. The word on the street was that the bishop was going to "smash Darwin."[7]

Is it any wonder that a group of serious biologists who had an important theory to discuss were disgusted and weary of this nonsense? Is it any wonder that Huxley himself wanted to avoid that debate because he was tired from arguing for several days already?

The famous debate was the culmination of multiple days of scientific presentations at the meeting. Two days earlier, Huxley had already been infuriated by a claim that gorilla brains and human brains were so completely different that humans and gorillas could not possibly be related. Huxley, an anatomist, knew better and said so.[8] Huxley was so exhausted by all the nonsense and arguing that he was ready to go home on Friday, the day before the great Saturday debate. He was talked out of leaving by one Robert Chambers.

Into this maelstrom walked Dr. John W. Draper, an American chemist, historian, and evolution enthusiast who had been born in England. His was the first paper on that Saturday morning and it must have seemed wonderful to be able to address such a packed house after years of laboring in what seemed like relative obscurity in the United States.[9] Although he accepted evolution, his paper wandered far from it, addressing instead his own ideas that not only were biological species subject to scientific laws but so were human physiology and the history and development of human civilizations. While physiology is indeed knowable through scientific investigation, the entire course of human civilizations is another matter entirely, and by the time Draper was done—over an hour later!—most people were bored and wanted to hear something more lively. They got something more lively.[10]

The Bishop of Oxford was eventually asked to speak, even though he was not a scientist and had never done any science. That's because this was back in the days when the thoughts of bishops were considered to be intellectually and socially important. The mass of clergy in the room loved it and made a point of laughing in all the right places, and cheered loudly and often.

Do you think this doesn't sound like church men, with their sober-sided reputations? Have you ever watched Joel Osteen preaching? Have you ever watched any evangelical preacher? Have you ever watched a crowd shouting "Amen!"? Preachers then knew how to whip up a crowd and appeal to emotions, just as they do today. The clerics in Oxford were banking on emotion carrying the day, so they were going to make the meeting as emotional as possible.

Although the bishop's remarks were not taken down at the time, they were the basis of a book review of *Origin* that he published— anonymously! —a few days later.

In that review, the bishop, Samuel Wilberforce, stated that he based his objections to evolution only on scientific rather than on religious grounds. But he lied. He lied in the way that denialists often lie, by mixing it in with so much overwhelming verbiage that it can be hard to keep track of when he's ignoring facts, when he's contradicting himself, and when he's simply getting things horribly wrong. His words are the very model of a smooth, genteel, highly articulate Victorian exercise of the "Gish Gallop." (In its modern version, the Gish Gallop is a debating technique used by the late creationist Duane Gish. It consisted of a rapid-fire, uninterrupted series of statements, arguments, connections, and conclusions on a wide variety of topics, that did not need to have any basis in fact or rationality.[11])

The review began by pretending to sound complimentary, talking about the excellence of Darwin's writing style. He says at the beginning that:

> all of these are told in his [Darwin's] own perspicuous language, and all thrown into picturesque combinations, and all sparkle with the colours of fancy and the lights of imagination.

Of course it is not necessarily a compliment to say that a serious scientific work contains "the colours of fancy"—which sounds like the author is a cute little school child—or "the lights of imagination"— which sounds like the author is an *imaginative* cute little school child. Imagination is primarily used to create works of fiction. This is not the way you describe the fifty year old author of a serious scientific work. Wilberforce went on to repeat various amusing natural history episodes that are related in *Origin*.

He made light of common descent by claiming that if Darwin could only prove it, he would accept his "cousinship with the mushrooms."

During the debate, he foolishly claimed that all men of science were hostile to evolution by natural selection, a claim that was provably false given the scientists who were present in that very room who were defending evolution. Of course this was also a way of belittling those defenders of evolution, claiming that they were somehow not real "men of science"—snobbish words, coming from a bishop who didn't do science at all! This claim memorably got slapped down later in the debate.

Then he got down to his own bad science.

First, he tried to make evolution fit into the church's model of the Great Chain of Being, pretending that all species were trying to move higher up this imagined scale going from plants to animals to humans, so that "turnips are tending to become men."

Second, he repeatedly confused natural selection with Lamarckism, known as "transmutation of species."

Third, he claimed that species never changed, and that proof of this could be found in the fact that the mummified remains of cats and dogs from ancient Egypt were exactly the same as modern cats and dogs, and this meant that evolution could not be true.

> From the early Egyptian habit of embalming, we know that for four thousand years at least the species of our own domestic animals, the cat, the dog, and others, has remained absolutely unaltered.

This meant that he was ignoring the work of geologists who had been saying for decades, with evidence, that the earth was far older than a few thousand years. Charles Lyell, the most distinguished of those geologists, was ridiculed by the bishop in this debate.

Fourth, he also ignored the vast amounts of fossil evidence showing that many species had begun, lived, and died out long before ancient Egypt ever came into being. He mentioned geology, and even different geologic periods, so he seemed to know something about them. But he "failed" to see, or chose not to mention, that ancient Egypt is geologically modern.

Fifth, he failed to understand that evolution does not generally occur by way of inter-species mating. Wilberforce said repeatedly and correctly that the offspring of inter-species matings, if they are produced at all, are generally sterile. But this has little to do with how most species originate through natural selection.

Sixth, he failed to understand that evolution by natural selection does what a good theory is supposed to do: point toward areas of research that might be fruitful. For instance, Wilberforce could not figure out how the electricity-producing organs in electric eels and other electric fish could have evolved.

> We see no possible solution on the Darwinian theory for the presence at once so marked and so exceptional of these organs.

He chided Darwin:

> for scarcely admitting that their presence is little else than destructive of his theory

Today, of course, having done research into physiology, we know that electricity is generated by both muscle tissue and nervous tissue. For instance, the heart is a muscle, and an EKG is a record of the electrical events that take place in your heart! We now know that the electric organs in these fish are evolved from nervous and muscle tissue. We also know that some fish emit weak electric currents that are used only for perception in murky water while others emit the killing levels of electrical discharge of electric eels. Thus it is clear how electric organs in fish were able to evolve.[12] But Wilberforce dismissed out of hand the idea that science could ever progress toward such new knowledge.

And then of course, in the end, he turned to religion, even while pretending not to. He said that Darwin:

> declares that he applies his scheme of the action of the principle of natural selection to MAN himself, as well as to the animals around him. Now, we must say at once, and openly, that such a notion is absolutely incompatible not only with single expressions in the word of God on that subject of natural science with which it is not immediately concerned, but, which in our judgment is of far more importance, with the whole representation of that moral and spiritual condition of man which is its proper subject-matter. Man's derived supremacy over the earth; man's power of articulate speech; man's gift of reason; man's free-will and responsibility; man's fall and man's redemption; the incarnation of the Eternal Son; the indwelling of the Eternal Spirit,—all are equally and utterly irreconcilable with the degrading notion of the brute origin of him who was created in the image of God, and redeemed by the Eternal Son assuming to himself his nature.[13]

Wilberforce continued to carry on about God for several more paragraphs, and after more windy nonsense eventually wound up his published review. The upshot of the whole thing is that after spending

many paragraphs pretending to talk about science, he ended up object-
ing to evolution by natural selection on religious grounds, just as one
might predict.

But at the debate itself, he threw in something extra, specifically to
annoy Huxley, who, as we recall, was already annoyed. Wilberforce is
said to have asked of Huxley: "Was it through his grandfather or his
grandmother that he claimed his descent from a monkey?"[14] This was
supposed to be a huge joke and personal insult.

Huxley, however, got up and tried to respond by talking about the
evidence for evolution. But people weren't listening. Unfortunately,
Huxley did not have a loud voice at the time and didn't know how to
carry an audience during a debate. So he gave up on patiently presenting evidence, and cut the bishop off at the knees instead.

Responding to the Bishop's comment about and apes and ancestry,
Huxley replied that:

> He was not ashamed to have a monkey for his ancestor;
> but he would be ashamed to be connected with a man who
> used great gifts to obscure the truth.[15]

Does this sound familiar? Could you be Huxley? How many of us
have been disgusted by denialists who use their talents to obscure reality rather than to illuminate it and face it? Scientists in general can
be patient, if dismayed, by the ordinary Christian fundamentalists we
meet today. But we are disgusted by professional denialists like Senator
Ted Cruz, who is clearly smart and educated, but who equally clearly
knows how to get money from the oil industry and votes from fundamentalists, and uses his talents to enrich himself and harm others by
refusing to publicly accept evolution, knowing it will cost him his base
of support, and outright denying the science of global warming.

Getting back to the debate—people who heard Huxley's dig were
shocked, because Wilberforce had religious privilege. He could insult
other people but other people weren't supposed to insult him back,
particularly not in public, in his own diocese. This may have been the
cause of the reported fainting.

Unfortunately, since Huxley did not know how to throw his voice at
the time, this means that although the bishop and some people nearby
heard his response, many others didn't. It was a packed noisy house in
the days before microphones were invented, so you couldn't be heard
unless you knew how to project your voice. In fact, on that day, Huxley

learned for himself the value of oratory and projection, and set out to learn how to do them, which made him the marvelous public speaker he later became.[16]

Soon after this, the *HMS Beagle*'s Captain Robert FitzRoy, who was also an invitee to this debate, joined the fray. His naval career had gone well, but his mind had become that of a Bible thumper. He brandished a Bible above his head and claimed that it was the source of all truth.[17]

Finally, the botanist Joseph Hooker spoke. Unlike Huxley, he could be heard above the din. When he addressed the audience, people listened. Hooker was a convinced evolutionist and already a distinguished scientist. He was a good friend of Darwin's but had only accepted evolution by natural selection because the evidence for it was overwhelming, based on Darwin's work and on his own research in botany. He talked about the evidence at length.

He made sure the audience knew that Wilberforce could not possibly have read the book that he was claiming to criticize. He also made sure the audience understood that Wilberforce knew absolutely nothing about botany.[18] The latter is very significant—an anti-Darwinian with great influence was a zoologist named Richard Owen, who had probably coached Wilberforce ahead of time for this debate. But Owen did not understand the *botanical* evidence for evolution, and neither did Wilberforce. As a result, Wilberforce tried to lie about the botanical evidence but was bad at it. (We will see later in this book how botanical evidence and botanists played a huge role in getting evolution accepted.)

Hooker finished by saying:

> I knew of this theory fifteen years ago. I was then entirely opposed to it; I argued against it again and again; but since then I have devoted myself unremittingly to natural history; in its pursuit I have travelled round the world. Facts in this science which before were inexplicable to me became one by one explained by this theory, and conviction has been thus gradually forced upon an *unwilling convert.*[19] [Italics added]

In other words, Hooker accepted evolution not because he wanted to but because, as an honest scientist, he had to. The bishop had earlier claimed that that no scientists accepted evolution. Hooker rejected this vehemently, personally, and publicly. But more importantly, he said why, and convinced at least some of the onlookers that he was correct.

After Hooker sat down the meeting disbanded, with all sides thinking they had won.

Both Wilberforce's ego and his position within the church would not allow him to admit that he had been bested by a bunch of socially inferior scientists. Huxley regretted the need to get personal with the bishop, but others were grateful that he had made the remark. Hooker's effective presentation of the evidence had probably helped the most.

But tellingly, the scientists had not been "smashed." In fact, they had come out a little ahead, especially in the realm of public opinion. Having public opinion on their side was what they needed, and the scientists reluctantly understood this.

Darwin himself wrote to Huxley the next month, saying:

> From all that I hear from several quarters, it seems that Oxford did the subject great good. It is of enormous importance to show the world that a few first-class men are not afraid of expressing their opinion. I see daily more and more plainly that my unaided Book would have done *absolutely* nothing.[20]

In other words, Darwin himself knew the value of his apostles—the people who risked their own careers, reputations, and peace of mind to help the idea of evolution by natural selection be accepted scientifically and publicly.

What's odd, however, is that evolution by natural selection had been presented to the British scientific world *two years earlier* and had not then caused great uproar.

The Famous Meeting before the Famous Debate

Consider the contrast: that raucous Oxford debate versus a real scientific meeting, the one where evolution by natural selection was *first* presented to British scientists. That took place on July 1, 1858, at a meeting of the Linnean Society of London.[21] At that scientific meeting, notes and essays written by two different scientists were presented, and both men proposed what we now call evolution by natural selection. All of these were read aloud to the group by Joseph Hooker.[22]

The first was some unpublished notes on evolution that Darwin had written down in 1839 and then copied and sent to Hooker and then Lyell in 1844. The second was a letter from Darwin to the Harvard botanist Asa Gray in 1857 that also outlined evolution. The third was an essay outlining evolution that had been written by Alfred Russel Wallace in February 1858, and sent to Darwin himself.

Together, this presentation was called "On the tendency of species to form varieties; and on the perpetuation of varieties and species by natural means of selection." It was published the following month in the *Journal of the Proceedings of the Linnean Society of London. Zoology.*

The meeting caused no great splash. The meeting was long. Other papers were read afterward on other subjects. Routine society business took place. The scientists listened, and some may have taken notes. The joint set of evolution papers was accepted for publication by the society's journal. No one fainted, no one cheered, and no one asked snarky questions about a person's grandparents. Our understanding of biology had changed forever, and almost no one outside the room noticed because it happened among a bunch of scientists.

Evolution was not fully accepted that day but it was also not rejected out of hand. This thoughtful response to such a radical idea may explain why scientific meetings today do not feature audience-happy bishops or people waving hankies.

Neither Darwin nor Wallace was present at that meeting. Instead, Joseph Hooker—one of Darwin's apostles—showed up and presented the work. Charles Lyell, an eminent geologist, also helped to actively ensure that the presentations were made.

Afterward Wallace went on to amass a huge collection of biological specimens in what is now Indonesia and nearby areas. Darwin went on to publish *Origin.* And then the fun began.

But why did all of these scientists —Huxley, Hooker, Draper, Gray, and Wallace himself—jump in to help promote this radical new idea? Why did these people risk their time, place in society, and professional reputations to defend someone else's work? Why was evolution by natural selection so important to them? Why did they all put themselves out for some other scientist? In short, who were these guys?

That is the story we will tell you in this book. But before we do, we have to explain where they came from, intellectually speaking. Who did they learn from? What was already known? What had already been discovered that helped them to understand that this far-reaching theory was correct?

How was it possible that evolution, that radical departure from religious dogma, was thought of at all, much less by two guys from the same country, independently, within a few years of each other?

In fact, it turns out that aspects of evolution by natural selection had been thought of by two *additional* guys from the same country, independently, within a few years of each other. This will be discussed in the next chapter. So—what foundations were already there that made this skyscraper of an intellectual achievement possible?

To explain evolution and Darwin's apostles, we must first tell you about their forebears. To understand Darwin's apostles, you must first learn about Darwin's scientific antecedents.

CHAPTER TWO
EVOLUTIONARY THOUGHT BEFORE DARWIN

"Nothing in biology makes sense except in the light of evolution."

Theodosius Dobzhansky (1900–1975)
Ukrainian Geneticist and Evolutionary Biologist

Darwin's Antecedents

Two hundred years before Darwin's time, in the 1600s, most people in England, Europe, and Europeans in North America believed the biblical account of creation. This includes the story of the Garden of Eden which describes how humans came into being along with all species of animals. In this account, the earth and everything in it was created in six days, and humans were made especially by God, in his own image, to rule over animals. People were told that the earth is only about six thousand years old. This supposed age of the earth was calculated by Bishop James Ussher in 1650 and was based on the number of human generations that there appeared to be in the Bible.

In the preceding centuries, Christian philosophers had also invented and described the existence of a Great Chain of Being. At the top of this hierarchical structure was God. Just below God came angels, followed in order by humans, animals, plants, precious minerals, and finally, dirt.

Fossils, which are found all over the earth, had had many interpretations in the preceding millennia. Christian philosophers sometimes said that fossils were organisms that had died in Noah's flood which were then washed away to unlikely places. This was used to account for fossil shells being found on top of mountains, for instance. Sometimes, however, fossils were thought to be geological formations that imitated life rather than biological organisms that had once been alive. But in general, by Darwin's time, what fossils are had only been established for about two hundred years, and they had only been subjected to systematic study for about fifty years.

However, cracks had already begun to show in the religious view of the world. By the time that Darwin set foot on the *HMS Beagle*, there were discoveries that had already been made, ideas that had already been suggested, and stubborn problems in religious thought that had already begun to disrupt this cozy view of how the world came into being. For this reason it is a good idea to set the stage and tell what scientists had already done; what people were thinking about in science, religion, and the world at large; and what problems they had when Darwin set sail.

The Great Tree of Life Instead of the Great Chain of Being: The Work of Carl Linnaeus
A Chart of Relatedness that Hints at Common Descent

The great tree of life, a huge branching diagram showing the relationships between all living things, was originally developed and brought into scientific use by Carl Linnaeus (1707-1778). We now know that this branching structure shows newer genetic lineages branching out from common ancestors, but at the time Linnaeus was simply being a careful scientist who classified things according to observable characteristics. These observable characteristics were meant to reflect natural relationships. Crucially, his branching system was based on the characteristics that organisms shared as well as the characteristics that were different.

Although Linnaeus did not grasp the idea of evolution by natural selection, he nonetheless set the stage for our later understanding of it by developing the tree-like system of classifying living things that is still in use today. The details of the structure have been changed over the years, as new information about plants and animals has been discovered, including information from DNA. However, the general branching structure, now known to reflect an evolutionary relationship between all living things, becomes ever more useful as time goes by.

In fact, it is so much a part of our thinking that it is hard to imagine doing biology without reference to it.

In making the tree of life, Linnaeus considered species to be real entities. But he also realized that they can be grouped together into genuses. Thus he invented the two-name (binomial) method of giving scientific designations to organisms, with an organism's genus being followed by its species. Thus, for instance, a blue jay which is in the genus Cyanocitta, and whose species designation is Cristata is, in scientific nomenclature, *Cyanocitta cristata*. However, in a blow to the religious idea that God created all species at the beginning of time, and that those species had not changed since, Linnaeus realized that plant species sometimes hybridize, and are therefore not fixed. Linnaeus also called nature a "butcher's block" and a "war of all against all," which shows that he realized that in nature there is a struggle for survival. But he considered that struggle to be a part of the divine order.

To understand how Linnaeus's hierarchical classification works, here's an example of how domestic cats are classified under Linnaeus's original system.

To begin with, cats are animals rather than being either plants or minerals. So cats go in the Animal Kingdom. Linnaeus divided the animal kingdom into six classes, one of which was Mammals. Cats are considered to be mammals because they suckle their young with milk and share other characteristics in common with all mammals.

In the original Linnean system, mammals are divided into eight divisions. Domestic cats go into the division called Ferae. All animals in the Ferae have cone-shaped teeth near the front of their jaws and pointy claws on their feet. Dogs, seals, weasels, skunks, and bears are also parts of this large group.

Ferae are then divided into genuses, with one genus belonging to all cats. This cat genus, called *Felis*, includes lions, tigers, leopards, jaguars, ocelots, lynxes, and domestic cats. Domestic cats are the species *Felis catus*.

What is interesting here is that within the genus *Felis*, all the members of this group really do look like cats to us. Most people are pretty comfortable with acknowledging that all its members look like cats. And they look like they are related, just as people who are siblings, cousins, second cousins, and even third cousins in any given family look like they are related.

The reason why people in a family look like they are related is because they share a common ancestor. Likewise, on a broader level, when we look at all the members of Linnaeus's group *Felis*, they look so similar to one another that we are pretty comfortable with thinking that all those members of *Felis* are related—that is, that they share a common ancestor.

So Linnaeus's system of classification hinted at the idea of common descent,—that is, it hinted at the idea of sharing a common ancestor. When Darwin wrote about the idea of common descent, people's minds had already been somewhat prepared for it, if they had read Linnaeus.

It is not a coincidence that Darwin and Wallace's ideas about evolution by natural selection were first presented at a meeting of the Linnean Society. Linnaeus was one of the preeminent European biological scientists of mid-to-late 1700s and most British, European, and American scientists who studied the biological world in any way had studied the Linnean system of classification. They were therefore already prepared, whether they knew it or not, to comprehend Darwin's idea of common descent.

The Question of Human Origins Begins Here, and So Does Religious Blowback

Linnaeus also made the ground-breaking decision to describe human beings just as he did any other biological organism. When he did this, he had to admit to the undeniable anatomical similarities between humans and other primates. He had done an anatomical examination of monkeys and found that the anatomy of monkeys was nearly the same as that of humans.

The only difference that he could rationally see was that humans could speak and other primates could not.[23] In fact, initially, he classified monkeys, apes, and humans into a group that he called "Anthropomorpha," which means "having human form."[24] Some scientists quibbled with this, saying that humans cannot be said to "have human form" since they are in fact humans. So he changed the title of the classification to "Primates." ("Prime" means "first" or "superior.")

There is great significance in the fact that he placed humans in the same group with monkeys and apes. In fact, he defied anyone to show him a meaningful natural difference between humans and other primates. Religious people, of course, objected strenuously. First of all, Linnaeus's classification of human beings with monkeys and apes meant that humans were animals. This meant that they could, if this were true, no longer be considered higher than all other animals in the Great Chain of Being. Second, this also meant that humans could not have been separately created by God in God's own image. Those religious objections continued to fester long after Linnaeus's time and created resistance to the idea of evolution by natural selection when Darwin published *Origin*.

Religiously-based objections to evolution continue to cause problems today, particularly because they have led to a degradation of science education in general and the understanding of evolution in particular. The ignoring of the clear predictions of evolution, whether through ignorance or because of apathy, has in turn led to the rise of predictable scourges like antibiotic-resistant bacteria.

Finally, from the religious perspective, there was the problem of the soul. From the Christian perspective humans were supposed to have souls, but mere animals did not. We will let Linnaeus have the last word on this matter:

One should not vent one's wrath on animals. Theology decrees that man has a soul and that the animals are mere 'aoutomata mechanica,' but I believe they would be better advised that animals have a soul and that the difference is of nobility. [*Diæta Naturalis* 1733]

Thus Linnaeus set the stage for the later development of the theory of evolution by natural selection. His classification system hinted at common descent, he realized that not all species are ancient, and he classified humans as animals, and more specifically as primates, along with apes and monkeys, despite religious objections.

The Age of the Earth—The Work of James Hutton and Charles Lyell

As noted previously in this chapter, a well-known Christian calculation of the age of the earth was done by the Irish archbishop James Ussher in 1650. This calculation was based in large part on trying to count the number of generations of people that are found in the Bible, and in part on trying to coordinate events in the Bible with the dates of known historical events. Bishop Ussher's calculation resulted in his announcing that the earth had been created in the evening, on October 22, in 4004 BCE. Other Christian scholars arrived at similar dates.

However, in 1785 a Scottish geologist, physician, and farmer named James Hutton presented research in Edenborough showing that the earth is ancient. Thus, only seventy-four years before *Origin* was published, people had not realized how old the earth really is. But the sparks of evidence were beginning to light a fire of fact about the matter.

Hutton used observations and predictions but no supernatural explanations in studying rock formations. He discovered two mind-boggling facts. First, rocks change. Rocks, that very symbol of a secure and unchanging environment, turned out to be going through constant changes, albeit at a very slow rate. Second, the fact that geological changes happened slowly led him to understand that huge amounts of time would have been necessary to make the geological formations that we can see with our own eyes. He realized, for instance, that sedimentary rocks were made of particles that had eroded off of other rocks, washed or blown away, gathered in one place (usually a lake or sea bed), and compressed. He realized, therefore, that thick sedimentary rock layers implied enormous amounts of time. He also realized that mountains rise and fall.[25]

Hutton first presented his research publicly in 1785. As is appropriate for any new scientific idea, however, acceptance was not immediate. There were religious objections as well, especially since many people, including other geologists, believed that present-day rock formations were the result of the biblical flood. But Hutton's ideas were eventually accepted by geologists. In 1830, the geologist Charles Lyell published the first volume of a three-volume book called *Principles of Geology*. The first volume rephrased and popularized Hutton's ideas and became the accepted text on up-to-date geological science as it was known at that time. This first volume was given to Darwin to take with him just before he set sail on the *Beagle*. [26]

Lyell's work proposed "Centers of Creation"—physical areas where new species came into being, fully formed. Darwin dutifully looked for them in South America but, ever the careful scientist, he had to admit that he never found them.

What he found instead were fossil sea shells on top of mountains, earthquakes that raised land such that he could witness it, and the multitudinous species, including those in the Galápagos, which slowly led him to develop the theory of evolution by natural selection. Seeing numerous species that were similar to one another but not the same, Darwin was able to formulate the idea that many species could evolve from a single ancestral species, given enough time. Armed with the knowledge of the immensity of time that Hutton had grasped and Lyell had explained, Darwin was able to see that evolution by natural selection was a workable idea, and the solution to the mystery of the origin of species.

The contribution that Hutton and Lyell had made to Darwin's thinking cannot be overestimated. These geologists showed that rocks change and that mountains rise and fall. They showed, with evidence, that the earth was much more than six thousand years old. That alone showed that the Bible is not literally true. But far more damaging was their other point: It is not just that the earth was created far more than six thousand years ago, it's that it is *still being created*. And it is being created by unconscious forces. So not only was there no single moment of creation, there was no conscious creator, either.

Therefore, the Garden of Eden story cannot be literally true. This means that the idea that the biblical story of creation is not literally true had already been broached by respected scientists, and widely accepted by scientists, long before Darwin wrote *Origin*. In fact, the knowledge that Darwin gained from reading Lyell's work profoundly affected how he interpreted what he saw while he was on the voyage of the *Beagle*.

This is a classic example of how advances in one field, in this case geology, can lead to unanticipated advances in other fields, in this case the development of the theory that unifies everything in biology: the theory of evolution by natural selection.

Fossils and the Problem of Extinction

Leonardo da Vinci (1452–1519) figured out that fossils were not only the remains of organisms that used to be alive, he also figured out that fossil shells on mountainsides were not deposited there by the biblical flood, since they weren't separated from one another and mixed in disorderly ways, as happens when areas are flooded, but rather showed the same sort of regularity that one sees in living beds of shelled creatures.[27]

Da Vinci, although best known today as a painter, did serious work in many fields of what is now called science. He did serious geological research, going into the mountains in northern Italy and studying rocks, fossils, and sedimentary deposits. He wrote down his conclusions but didn't publish them. These studies have been found in Leonardo's notes by later scholars. Since he did not publish his conclusions about fossils, it is unlikely that Darwin knew specifically about Leonardo's remarkable insights.

The Englishman Robert Hooke (1635–1703), who made discoveries in many scientific fields, including geology and biology, figured out that fossils are the remains of once-living organisms. He also figured out that some species of those once-living organisms were no longer in existence, that is that those species had gone extinct. Additionally, he figured out (as had Da Vinci) that fossil seashell beds that exist in areas that are now dry simply show that the dry area was once under water—that is, that not only do organisms go extinct but that watery areas can rise in altitude and become dry, and as a result, whole ecosystems can change. Unlike Da Vinci, Hooke published his findings, with his work on the subject, *Discourse of Earthquakes*, being published two years after his death.[28]

The French scientist Georges Cuvier (1769-1832) studied fossils in a systematic way. He understood that they had once been living organisms and realized that some of those once-living species had become extinct. He made it a mission to have the world accept extinction as a fact, despite religious objections.

Extinction is a problem for a biblical view of life since it would seem strange for God in his perfection to create perfect organisms only to let them die out. This compromises the idea of God as an all-knowing and all-wise creator. Cuvier is known as the "father of catastrophism," which is a concept used in his time, and today by creationists, which allows for events like Noah's flood and other biblical disasters while placing them into a scientific context, thus merging faith and science.[29]

Cuvier, however, demonstrated that some fossil organisms were simultaneously clearly related to living species but not the same as living species. For instance, he established that African elephants and Indian elephants are living species that are distinct from one another, and also showed that extinct European and Siberian mammoths, though similar to living elephant species, were a different species, and extinct.

More generally, he also made it clear that fossil organisms are related in various ways to extant organisms, and he included them in his general family tree of life.

William Smith (1769-1839) was an English surveyor who became involved in the creation of the Somerset Canal. This required him to survey possible routes for the canal, and this in turn required him to study the local rocks through which the canal would be dug.

When he engaged in that study, he found there were fossils in the sedimentary rock. As he continued to study them, he found that different layers of sedimentary rock had different types of fossils in them. Even more important, he found that the different layers of fossil deposits always occurred in the same order, from deepest to shallowest.

What's more, he found that he could predict which layers of fossils would appear in which types of rock, even when the rock beds were not near to one another. He also found that groups of fossils co-occurred. In modern terms, what he was seeing was the remains not just of individual organisms but of ecosystems of organisms that lived together.

The fact that these groups of fossils co-occurred in the same order every time became known as "faunal succession."[30]

Many questions arose as scientists began to understand fossils better. As fossils were found and studied and classified, it became clear that many, even most, fossils were of extinct species. This led to the question of where all the fossils of current-day animals and plants are. If all these fossils of animals that no longer exist are being found, and all species everywhere were created by God at the same time, then why aren't fossils of current-day plants and animals being found as well?

Even more importantly, as noted earlier, extinction is a problem for those who believe in biblical creation. If God created all organisms at the time of the Garden of Eden, and if God is perfect, then how could God create a species that would die out and become extinct? Why would God do that? And why did God do that so many times?

It became reasonable to wonder: Why do species go extinct? How does this fit into God's plan? And is there another explanation?

The Struggle for Survival—The Work of Thomas Malthus (1766-1834)

In 1798, Thomas Malthus published *An Essay on the Principle of Population*, a treatise that explored the relationship between the availability of food and population numbers in humans. What he found was that when more food became available, people were better off for a while, but that eventually the population grew and used up all the available new food.

Basically, he found that people reproduced pretty rapidly when more food became available—and while there was enough food, people were healthy, but when the population numbers outstripped the amount of food available, then more children died young, more people starved to death, and more people who were weakened by malnutrition died of disease. According to Malthus, the population would eventually stabilize again at a level at which all available food was being eaten, and poor people struggled for existence and were hungry.[31]

Darwin read this essay and did what insightful people do. He generalized from Malthus's example of human populations to seeing a struggle for existence that goes on all the time among plants, animals, and all biological organisms. This is a driving force in natural selection. Darwin wrote about this in *Origin*:

> As many more individuals of each species are born than can possibly survive; and as, consequently, there is a frequently recurring struggle for existence, it follows that any being, if it vary however slightly in any manner profitable to itself, under the complex and sometimes varying conditions of life, will have a better chance of surviving, and thus be naturally selected. From the strong principle of inheritance, any selected variety will tend to propagate its new and modified form.[32]

His insights about this struggle are due in part to his having read Malthus's work. Wallace, Darwin's co-discoverer of evolution by natural selection, also read Malthus's work, was influenced by it, and credited Malthus's essay with helping him to formulate his ideas on evolution.

Suggestions of evolution (but not natural selection) and suggestions of both natural selection and common descent during Darwin's time.

The idea that biological species evolve, that is, that they change over time, had been proposed long before Darwin, and more than once. So it was not an unheard of idea. However, it had not been accepted by the scientific community nor by the general public and certainly not by the various Christian churches. Moreover, no one had come up with the mechanism of natural selection, which is the process that correctly explains how evolution can take place. The theory of evolution *by natural selection* is the matchless contribution made by Darwin and Wallace.

Suggestions of Evolution and the Idea of the "Transmutation of Species"

France

In France, the prolific naturalist Georges-Louis Leclerc, Comte de Buffon (1707-1788) published a monumental forty-four volume encyclopedia of all that was known about the natural world during his time. In this work, he discussed the similarities between humans and apes and suggested that humans and apes might have common ancestors. He also believed that organisms change but did not suggest a mechanism. In a separate published work he also suggested that the earth is much older than the six thousand years that had been proclaimed by the church. Thus he managed to counter nearly two thousand years of dogma, opposing the ideas of special creation and the Great Chain of Being on the one hand, and the six thousand year old earth on the other.[33]

Later, Jean-Baptiste Lamarck (1744-1829) also suggested that species change over time, that is, they evolve. However, the mechanism that he suggested, the inheritance of acquired characteristics, is now known to be incorrect. He called this suggested process the "transmutation of species." [34]

His work was derided at the time, but it is nonetheless the case that he published his ideas and further promoted the general idea of biological species changing over time. That is, he introduced the concept of evolution but not the concept of evolution by natural selection. Unfortunately, the idea of transmutation of species was so thoroughly rejected by scientists and churches alike that it actually caused some problems for Darwin and Wallace in getting their own, correct, ideas to be accepted.

England

Darwin and Wallace also had some influences that were closer to home. In Darwin's case, particularly close to home. Charles Darwin's own grandfather, Erasmus Darwin, had suggested that biological species evolve, but again did not envision the correct mechanism.

Erasmus Darwin (1731–1802)

Charles Darwin's grandfather was an avid naturalist as well as a physician. He had ideas about evolution, in the sense that he thought that organisms change over time, and he evidently agreed with Hutton that time had been long. His best-known poem on the subject, *Zoonomia,* stated:

> Would it be too bold to imagine, that in the great length of time, since the earth began to exist, perhaps millions of ages before the commencement of the history of mankind, would it be too bold to imagine, that all warm-blooded animals have arisen from one living filament, which THE GREAT FIRST CAUSE endued with animality, with the power of acquiring new parts, attended with new propensities, directed by irritations, sensations, volitions, and associations; and thus possessing the faculty of continuing to improve by its own inherent activity, and of delivering down those improvements by generation to its posterity, world without end![35]

This tells us that in this poem his evolution is similar to that described by Lamarck. But in a small way he anticipated natural selection as well, stating in this poem that "three great objects of desire" for every organism were "lust, hunger, and security," and that "the strongest and most active animal should propagate the species, which should thence become improved."

Although he died before Charles was born, his life's interests are reflective of those his grandson would adopt as well. He was an avid botanist and formed the three-member Lichfield Botanical Society for the express purpose of having Linnaeus's plant classification scheme translated into English. He also opposed the slave trade and was a member of the Lunar Society of Birmingham, a dinner club and society of learned men which met to discuss science, invention, and other matters of importance. This anticipates the X-Club that was formed in Charles Darwin's time in order to advance the acceptance of evolution by natural selection and to advocate for evidence-based professional science in general.

Robert Chambers (1802–1871)

A publisher and amateur scientist, Chambers wrote the 1844 book *Vestiges of the Natural History of Creation* and published it anonymously. This book argued that both stars and biological organisms change over time. The science in it was pretty bad but it was a runaway hit. Prince Albert is said to have read it aloud to Queen Victoria. The type of biological evolution he suggested was not natural selection but instead something closer to Lamarck's version of evolution, which most scientists had discredited at that time. In any case, it created a storm of controversy simply because the subject matter itself suggested that God did not actively maintain both social hierarchies and organismal ones. [36]

This flawed work was rejected by most scientists at the time but it did introduce the idea of evolution to popular culture, and it thus may have prepared people to be more accepting of the idea once *Origin* was published.

One Clear Announcement of Evolution by Natural Selection that Went Unnoticed: Patrick Matthew

In 1831, a book called *Naval Timber and Arboriculture* suggested natural selection.

> There is a law universal in nature, tending to render every reproductive being the best possible suited to its condition that its kind, or that organized matter, is susceptible of, which appears intended to model the physical and mental or instinctive powers, to their highest perfection, and to continue them so. The law sustains the lion in his strength, the hare in her swiftness, and the fox in his wiles. As Nature, in all her modifications of life, has a power of increase far beyond what is needed to supply the place of what falls by Times' decay, those individuals who possess not the requisite strength, swiftness, hardihood, or cunning, fall prematurely without reproducing—either a prey to their natural devourers, or sinking under disease, generally induced by want of nourishment, their place being occupied by the more perfect of their own kind, who are pressing on the means of subsistence.[37]

This passage clearly indicates natural selection. It also, as with other thinkers about evolution, suggests the influence of Malthus. But while the article is remarkable and quietly audacious, it only discusses the winnowing aspect of natural selection, which is the part that is easier to understand. The slow formation of multiple new species out of one existing one, that is the *creative* aspect of evolution known as common descent, was not suggested in this article.

The author, Patrick Matthew, didn't have the mountains of data that Darwin amassed. In fact, Matthew admitted that this idea was more of an insight than something he had arrived at through careful research.

That said, when Matthew wrote to Darwin after *Origin* had been published, and complained that Darwin had not cited Matthew's earlier work, Darwin, ever the gentleman, wrote back and apologized, saying mostly that he hadn't been able to read *every* article in every journal prior to publishing *Origin*. He also appropriately cited Matthew in later editions of the book.

This episode does however show the influence of class in Victorian England. Matthew's article and book, though widely reviewed, did not cause a stir in either scientific or religious circles. This may have been partly because of its less far-reaching conclusions that did not involve describing how new species came into being. But another factor was probably the fact that farming (arboriculture) and naval timber were just not topics that excited the notice of many Victorian scientists, nor did they threaten the primacy of organized religion as a source of knowledge.

To his credit, Darwin did consult farmers and breeders in the course of his research for *Origin*. Had he consulted with timber growers instead of pigeon fanciers, he might well have conversed with Matthew himself, and cited him properly the first time.

And a Suggestion of Common Descent from a Colleague of Darwin's

Edward Blyth was a contemporary of Darwin's who spent most of his working life in India. A zoologist, Blyth wrote several articles for *The Magazine of Natural History and Journal of Zoology, Botany, Mineralogy, Geology and Meteorology* during the 1830s. One in particular, from 1837, stands out as a precursor to Darwin's thoughts:

> It is a positive fact, for example, that the nestling pluage of larkes, hatched in a red gravelly locality, is of a paler and more rugous tint than in those bred upon dark soil. *May not, then, a large proportion of what are considered species have descended from a common parentage?*[38]

Darwin refers to Blyth's work in his essays of 1842 and 1844. They corresponded in the 1850s, and Darwin refers to Blyth in the first chapter of the first edition of *Origin*. So in this case it is fair to say that Blyth's research and thinking positively affected Darwin's development of the idea of evolution by natural selection.

Unanswered Problems in Religious Thought.

There were also some religious problems having to do with biology that were not adequately answered by the Bible or by theology. These too played a role in weakening the comfortable view that God had created all life on earth in six days roughly six thousand years earlier, at the beginning of biblical time.

The Problem of Pain and Suffering

The Christianity that Darwin had learned said that pain and suffering somehow fit into God's overall plan, and that the overall result was positive. But Darwin couldn't see the divinity in some of the adaptations that he had studied. For example, consider the ichneumon (Ichneumonidae) wasps (of which there are over eighty thousand species!) that paralyze caterpillars and lay their eggs on them. Once the eggs hatch, the live but paralyzed caterpillar is eaten by the larvae that hatch out of the eggs.

Darwin wrote:

> I own that I cannot see as plainly as others do, and as I should wish to do, evidence of design and beneficence on all sides of us. There seems to me too much misery in the world. I cannot persuade myself that a beneficent and omnipotent God would have designedly created the *Ichneumonidae* with the express intention of their feeding within the living bodies of caterpillars, or that a cat should play with mice.[39] [Letter to Asa Gray, May 22, 1860]

The Problem of Numbers of Species

A simpler problem exists around the sheer number of species or organisms that exist in the world. Darwin had seen many different types of organisms, and numberless varieties of them. How could all of those species have fit onto Noah's Ark?

The Problem of Extinction

This problem did not exist in earlier centuries of Christian thought, but once fossils were purposefully dug out and classified it became apparent that many species had lived for a long while on earth and then died out.

This brings up two troublesome questions. First, if all animals were made during the six days of Genesis, then why aren't extinct species mentioned as existing in the story of the Garden of Eden? For instance, where are the dinosaurs? Animals that magnificent and numerous and varied ought to have had a mention if they were made by the creator and lived at the dawn of time in the Garden of Eden. But the Bible is silent on the subject of dinosaurs, pterosaurs, ichthyosaurs, and other massive and fascinating animals.

This is of course why the Creation Museum in Kentucky has made a point of concocting the story that, really, there *were* dinosaurs in the Garden of Eden. The Creation Museum has displays showing them there! This may satisfy children who might be lured away from conservative Christianity if it meant not getting to love dinosaurs anymore, but it doesn't address the greater problem, which is this: what about *all* the species that have gone extinct? Would not wooly mammoths have deserved a mention in the Bible? And giant birds like moas? Why didn't all those fishermen in the Bible pull up the occasional trilobite? Trilobites were the most common type of organism on earth for millions of years!

Isn't it odd that the species mentioned in the Bible are ones that still exist in the modern Middle East?

Second, there is the aforementioned problem of God's "perfect creation." If all these extinct species were made by God, why did he make them only to let them die out? If they were imperfect, then why did he make them in the first place? Surely the creator of the universe could do better! And if they were perfect, then why did they go extinct?

The Problem of Today's Organisms and the Fossil Record

A related problem has to do with today's organisms. None of these, including humans, are found in the fossil record. If all species really had been made at the beginning of the world, then we would expect to see dinosaurs, trilobites, saber-toothed cats, modern humans, and all their domestic animals such as modern horses and modern cows, all present in the fossil record, all in the same layers. This is not what is ever seen, and this causes problems for those who want to believe in the biblical story of creation.

These problems may have helped to pry people, including Darwin, away from a literal interpretation of the Bible and away from the idea that there is an all-powerful, all-loving God who is in charge of everything.

A Word on the Structure of This Particular Scientific Revolution

Evolution by natural selection might appear to fit right in with Thomas S. Kuhn's idea about scientific revolutions being the products of human enthusiasm and culture.[40]

But as our section on Darwin's antecedents shows, vast amounts of scientific work had already been done that cleared the way for scientists to be able to think of evolution by natural selection. It is clear that all the people who thought of evolution by natural selection—Darwin, Wallace, and to an extent Matthew and Blyth—were standing on the shoulders of giants like Linnaeus and Cuvier.

It is also crucial to realize that the reason why most of the apostles came to accept evolution was not because they particularly wanted to accept it but because Darwin's mountains of data gave them no other choice. Hooker described himself as an unwilling convert and Asa Gray fell into that camp as well—but both were won over in large part because of Darwin's exhaustive work, and because the theory explained things that were otherwise unexplainable. Two of the apostles, Gray and Wallace, both had intractable mysteries in their own research that were solved by evolution by natural selection. Once they were on board, the apostles threw their social and scientific weight behind getting evolution accepted. But they didn't sign on because they liked Darwin or particularly wanted evolution to be true—it was because they had to admit that it *is* true.

The men described in this book worked very hard to get this revolution in biological thought accepted. They weren't convinced because it was fashionable; they were convinced because of mountains of data. Then they worked hard to *make* the idea fashionable. Likewise, the idea of evolution by natural selection didn't come out of nowhere, nor did it arise because Darwin was particularly bold. It came out of years and even centuries of work by others, which made the idea plausible and possible to envision.

The only thing that made evolution shocking was not that it contradicted a *scientific* paradigm but that it contradicted a religious book. That is still the only thing that, to some, makes it shocking today.

So again we ask why the apostles did what they did. Why did they risk their time, energy, and fortunes to promote and protect a theory when the immediate consequence of doing so was blowback from powerful religious institutions and the public opinion that those institutions commanded with a heavy hand? Why did the apostles care so much? In short—who were these guys? They risked a great deal and deserve to have their stories told.

CHAPTER 3

DARWIN AND DEFINING HIS APOSTLES

"I do not feel obliged to believe that the same God who has endowed us with sense, reason, and intellect has intended us to forgo their use."

Galileo Galilei (1564–1642), Astronomer

This book is primarily about the importance of Charles Darwin's groundbreaking work that has underscored and framed the science of biology for the last 170 years. Darwin's theories impact how much we know about ourselves as well how much we know about every other existing and extinct species that has swum, crawled, walked, leaped, and flown on our planet.

The book is intended to frame Darwin at the center of a revolution, where he is the sun in a solar system surrounded by people who shared and basked in the light of his giant ideas and his theories about nature.

The planets supporting and surrounding our sun called "Darwin" are themselves esteemed. They include eminent scientists and advocates for natural selection like American biologist Asa Gray, the British immigrant to the United States John William Draper, as well as Darwin's British colleagues Joseph Dalton Hooker, Thomas Henry Huxley, and Alfred Russel Wallace. These men were not just satellites orbiting Darwin but leading scientists of their day who played a crucial role as Darwin's advocates in British, continental European, and American scientific and cultural circles far away from Darwin's fortress at Down House in the English countryside.

The Five Apostles: Draper; Gray; Hooker; Huxley, and Wallace

When imagining this book we spent many hours thinking about Darwin the man, both in terms of his brilliance and his many foibles. A deeply intelligent and insightful lone wolf who in the end needed others to spread his ideas and who relied on less than a dozen immediate colleagues to bring truth to science. We even reached out to local clergy for guidance (not to repent but to discuss the book's title) to be sure that "Apostle" in the context of the book's purpose would be the right word to convey the meaning of how Darwin's friendships saved his ideas from obscurity.

We can recognize the irony of Darwin even in our own actions. His was an all too human life and one that in many ways matches our own. Like in our own lives, Darwin had opportunities and challenges thrust upon him. Sometimes he gained from them and sometimes he lost. He could not know that choices he made early in life would set him on a new course; and later in life he would also feel the penetrating loss of loved ones to illness and disease.

In this way we should remember that every great mind is still a human mind encased in the confines of a human body. We live a linear life without the benefit of a crystal ball. Darwin need not be deified as a perfect, all-knowing being—regardless of his deep insights into the processes of biology and evolution.

Charles Darwin was a brilliant semi-recluse who made his home at Down House, now the Darwin Museum, which served as an unlikely place to foment a revolution in the human understanding of nature. This is part of the paradox of Darwin's public and private lives.

Had Darwin not been surrounded by equally brilliant colleagues and friends who not only pushed him to publish but also took the time to advocate for his ideas, we may not have the quality of life or the science we have today. Just as the Beatles would sing, "I get by with a little help from my friends," more than one hundred years after the publication of *Origin*, so too did Darwin depend on his friends to be his advocates in great debates which he could not attend. His colleagues fended off the antiscience and religious intellectual executioners whose unifying goal was to see Darwin's science ridiculed and rendered extinct.

Certainly it took more than Darwin to free our collective minds for a human future in which we all continue to benefit—thanks to advances in biology, chemistry, and so many allied hard and soft sciences. It took a village to make evolution understandable and accessible to the masses. This village may have been founded by Darwin but it was inhabited and managed by Huxley, Gray, Hooker, Wallace, and Draper as well as many others throughout the scientific world.

Perhaps it is audacious to write, but we could easily make the case that *Origin,* which sold out on its first day of publication, is a more profound and moving book than many of the holiest books ever written. Holy books require the acceptance of specific religious or spiritual doctrines to make their content and claims relevant. Therefore one's personal holy book could be discounted and even ignored by others who have chosen different faiths around which to build their individual identities and communal cultures. This makes all religious doctrines subjective rather than objective, which is not the case with Darwin's work.

The theory of natural selection found in *Origin* is objective truth, just as the theory of gravity and the germ theory of disease are also objective truths in science. While we are all entitled to our opinions, we aren't entitled to different concrete material facts. And everything evidenced in natural selection is indeed a fact.

For Darwin's five colleagues, there was a symbiosis to being in league with their friend and in many cases their mentor. Being part of Darwin's research and life, being his companion and confidant while also having masterful debating skills, having research journeys of their own to draw from while being accomplished scientists in their own right, offered each apostle entré to the highest and most respected scientific societies in Europe and the United States. For Darwin the explorer, naturalist, and sage it was fortuitous in many ways to have friends who would essentially come to his defense, helping him claim his right as "discoverer" of the most vital and misunderstood natural processes found in nature.

Draper, Gray, Hooker, Huxley, and Wallace were indeed Darwin's scientific contemporaries and each played an important role during the heady early years in shaping and winning the early debates concerning natural selection. But these men were more than advocates. In some cases they formed a tight circle around Darwin, urged him to publish, and pushed his ideas into the mainstream. They were in fact Darwin's apostles.

What is an Apostle?

While in modern times "apostle" rings to an immediate religious connotation, the word in reality is older than the faith traditions that commonly claim it to define the original followers of some very specific prophets. The *Oxford English Dictionary*, the source for word etymology, provides the best original meaning. Greek in origin, the word sim-

ply means "messenger" or "a person on a specific errand"[41] In the last
two thousand years and certainly within the codified story of Jesus
Christ, the word has become part of the Christian vernacular to define the twelve men who served and studied under him. Thus the word
seems forever linked to a religious rather than a secular meaning.

Based on the original meaning of the Greek word it would appear that
both religious and secular apostles are typically advocates who spend
their time teaching others the good news of their views. In this way the
five men who would help Darwin usher in modernity and the positivity
that comes with understanding natural selection share the same traits as
the storied twelve men who chose to study the teachings of Jesus Christ
and then speak on his behalf after his crucifixion and resurrection.

Perhaps the main difference between Darwin and Christ's apostles
is one of evidence and verifiability. We have little historical evidence
other than scripture to tell us about Jesus or the deeds of his twelve
apostles. These men we are told took the murdered prophet's teaching
from memory to the towns of the Roman world. It is said that Jesus
only preached and never wrote anything. Such hearsay has a long theological tradition. The Christian Scriptures (commonly called the New
Testament), the Hebrew Bible (the Torah), and the Qur'an are each an
oral tradition and essentially a group project. All have been edited, re-
interpreted, and redefined over the ages rather than being some unal-
tered world of God, as believers would like us to imagine.

In Darwin's case, his five apostles could benefit in ways much greater than the ones who followed Jesus. Darwin lived another twenty-three years after he published *Origin*, and in fact he wrote numerous
other books—eleven total—about nature and the role natural selection plays in the evolution and extinction of plant and animal species.
These books include important titles such as *The Descent of Man* and
The Expression of Emotions in Man and Animals.

From the perspective of Darwin's apostles, having your mentor
around for an additional two decades after he published his germinal
work allows you time to ask questions and develop your own work.
This full and direct experience can bolster your advocacy and tenacity, making you even more determined to see his ideas sprout, grow,
and then become accepted. What's more, these men carried on their
advocacy even after Darwin's death. Of these five men who surrounded Darwin, two lived more than a decade into the twentieth century.
Joseph Hooker lived until 1911 and Alfred Wallace until 1913. And
they protected Darwin's ideas throughout their lives, hoping that fu-

ture generations would benefit.

Darwin's theory of evolution was revolutionary. However, any time there is a revolution, be it social, political, or intellectual, there will be winners and losers. In the case of evolution it was both science and our modernity that were and still are the winners.

If Wallace hadn't provided the competition that Darwin needed to finally publish his ideas about evolution, it is likely that others eventually would have.

In a global sense, the western world was finally ready for a complete synthesis of the good ideas and observations that led to the development of the theory of natural selection, and it was likewise ready to reject creationist ideas that had both limited and hobbled science and scientists over the generations. Although it was Darwin who first crossed the line into modernity, it was really only a matter of time until some member of the scientific community "discovered" what had existed since the dawn of life itself.

The mechanics of both science and the scientific method had themselves evolved and matured since the European Enlightenment. In 1859 the time was simply right for an evidence-based theory of evolution. Darwin's precision along with his observations and elegant writing made him the man who would change the world. For this we owe him and those who supported him our admiration. They saved biology and our science from a strict theocratic view.

Of Truth and Truthiness

Although late night commentators like Stephen Colbert sometimes joke about the "truthiness of ideas" we know from the scientific perspective, which is based on evidence, observation, and the verification of findings, that the process of evolution is indeed true. Natural selection is true whether you accept it or not. It doesn't need belief to make it work. Evolution is a conclusion based on facts and not a philosophical construct based on opinion or the earnestness of the believer.

The hostility today shown by those who deny or doubt evolution, now over 150 years since the publication of *Origin*, is equal to the hostility that the original advocates of natural selection encountered. Such hostility towards scientific discovery demands our resistance. The freedom to make scientific inquiry and produce results—and to then share ideas, theories, and conclusions—all these remain vital to scientific investigation and must be protected. If we do this, then the commerce so often created by scientific discoveries, and the academic study it always improves, will also be protected.

We'd be foolish to not notice the rabid denial on religious grounds of Darwin's theory in his day. Evolution by natural selection presents an inherent conflict with theology because it dethrones humans by placing humanity in with the rest of nature. It does this while simultaneously denying special creation—the fundamental creationist idea that all life in the universe was created through miraculous processes by a supernatural force—on the grounds that there is no evidence for it.

In the United States more than two-thirds of the population believes that evolution is false or that it was directed.[42] Throughout U.S. history there have been legal dustups by religious groups in an effort to stop the teaching of evolution. In 1925 there was the infamous Scopes Monkey Trial. Eighty years later in Dover, Pennsylvania, there was another failed legal battle to force "intelligent design" into public high school science classes. In the state of Texas and in many states where there are proponents of theology-based creationism via intelligent design, there are frequent attempts to change curriculum and textbooks to water down evolution or infuse theology into biology classrooms and curriculum. As of this writing, the states of Louisiana and Tennessee allow creationism/intelligent design to be taught in public schools as science.[43]

As mentioned above, there is even a creation museum in Kentucky that offers museum goers exhibits that show velociraptors gingerly walking alongside people as if these vicious dinosaurs were pets. This isn't real science. This is Flintstones science. This is delusion. But it is exactly how those who believe in special creation must bend reality into the shape of a pretzel to accommodate things that are clearly evidence for evolution and an ancient earth—like dinosaurs—into their biblical worldview.

In 2017, the Museum of the Bible opened in Washington, DC.[44] Like the creation museum, this facility lauds the Bible and the mythology of Judeo-Christian biblical creation. The funding for the institution comes from private wealth, and while the goal is to serve as an educational center, it in fact serves as another physical place where believers can go to verify what they already do not challenge, that life is due to supernatural forces.

For these real or impending affronts to evolutionary science, which tell a very different narrative, we must all continue to stand by and for reason and the natural sciences. Evolution by means of natural selection remains one of the most liberating, influential, and important scientific theories in human history. Charles Darwin's name can typically be found up near the top of lists that codify the great thinkers in history. Natural selection rivals similar theories of importance such as gravity, germ theory, and $E=mc^2$.

The theory of evolution has direct impact on the operational knowledge of how life works on earth. And based on rational study, it has implications on how we think organic life may work on other worlds and throughout the cosmos. Darwin's theory has led to discoveries in numerous areas of scientific inquiry, both in and outside of biology. This includes chemistry, geology, physics, medicine, ecology, computer science, anthropology, and astronomy. Research in these and other areas and the allied discoveries that come from scientific inquiry help us not only to understand our humanity and our big-brained primate responsibilities to each other but how these responsibilities transfer to other species and our planet.

Darwin, the Apostles, and Biology's Impact on Our Culture

"Let us hope it is not true, but if it is, let us pray it does not become widely known."[45] This statement about evolution via natural selection has been attributed to several people, most usually to either the Bishop of Birmingham or the Bishop of Worcester's wife. Even if attribution has never been finally established, the spirit of the quote pretty much sums up how those vested in biblical versions of creation felt, and in many cases still feel, about biological evolution.

A quote that can actually be attributed to a real person best describes the importance of natural selection. It was written by Theodosius Dobzhansky, a Russian Orthodox evolutionary biologist who wrote in his 1973 article in the *American Biology Teacher* that "Nothing in biology makes sense except in the light of evolution."[46] This was Dobzhansky's clear shot at past and present biblical creationists in their efforts to contort science to prove their theology correct and at the same time disprove natural selection.

As a result of Darwin's research and writings, the field of biology has grown considerably, both as an academic pursuit and as a field of study which provides real-world solutions that touch everyone. People from farmers to doctors, pharmacologists to food scientists, space explorers to oceanographers, and consumers and citizens of every nation have all benefitted from our understanding evolution. These benefits have even accrued to people who don't accept the science.

While Darwin had unlocked and formally labeled the natural process of biological change over time through evolution, he did not have knowledge of genetics—the critical mechanism through which evolution operates. For that part of the story of science, we have to look to a nineteenth century Augustinian friar named Gregor Mendel.

Mendel's work in genetics brought organized scientific investigation to the process of breeding plants and animals for agriculture. Humans had been breeding plants and animals for ten thousand years but the laws of basic genetics were not known until Mendel discovered them. Mendel's laws of inheritance are fundamental to understanding how genetic traits are passed from one generation to the next, whether the selection process is done by a breeder with a particular outcome in mind or by natural selection which never has an agenda.

Although Mendel and Darwin were contemporaries, and Mendel published the results of his plant experiments in 1866, his work was not widely known. It was only in the early 1900s that biologists rediscovered Mendel's work and its significance was understood. The combining of Mendel's genetics with Darwin's theory of natural selection is what has given modern biology its power.

We now understand the inheritance of traits and how species either adapt or become extinct through evolution. As we have become more sophisticated in our understanding of both genetics and evolution, whole fields of science have come into being. These provide frameworks for still further human exploration of the natural world.

Today it is estimated that there are almost three hundred thousand biologists in the field or in allied fields in United States alone.[47] Cell biologists, plant biologists, geneticists, biochemists, experimental biologists, marine biologists, bioengineers, astrobiologists, and epidemiologists are just some of the diverse tracks one can specialize in. Not to mention bioethicists and biology educators who teach the next generation of discoverers and "uncoverers" of nature's hidden realms and how to be good and rational science professionals.

What aspect of our modern world culture hasn't been touched in some way by modern biology? From the foods we eat to the medicines and medical treatments we receive, to new fields of research such as the expansion of hybrid computer-biological devices for information storage and retrieval, we cannot walk down the street of any small town, let alone a busy metropolis, and not see, smell, hear, taste, or feel how the current field of biology has impacted our very modernity.

We are surrounded by evolution and it is up to science and its advocates to remind us fully of this fact. It is inside us working on the genetic level to shape our organs and longevity; it is operating on every species that we have identified and is also working on species yet to be found and named taxonomically.

Evolution is also outside us and can be viewed in the museums and zoos, parks and mountain trails, ocean shores and supermarkets we may visit over the course of our lifetimes. Denying evolution is in effect denying life itself. It is the conscious choice to remain unaware and ignorant about the universe and how the handiwork of nature operates—even without our full consent or full understanding.

Conversely, accepting the natural world, as biologists and as many lay people do, one can gain a grander understanding of how evolution and natural selection shape not only our humanity but also every other living (and extinct) creature on our planet. Once you accept evolution as true you cannot help but gain an immediate respect for the fundamental laws of nature. Once you accept evolution you cannot help but gain an immediate respect for the plight of the earth's species, including our own: how much by chance we all got here, how fragile we are and our planet is, and how we are interconnected and interdependent on each other for survival.

Biologists can explain with deep evidence how evolution is an ongoing process—one that can never stop since, as an unconscious and natural process, it is playing out within every species and ecosystem. This interdependence within nature has succeeded for billions of years. It is known succinctly, thanks to Australian physicist Fritjof Capra, as "the web of life."[48]

This is the unbiased work of evolution. Had our ancestors not survived when our species was a relatively small gathering in Africa, we would not be here to write this book and you wouldn't be here to read it. That's because our parents would not have been here, nor their parents or their parents before them. Whether one studies biology or not, we should all find our existence remarkable, but also wonderfully satisfying. In the lottery of the natural world, the fact that we exist at all makes us all very special.

Human Made Adaptations in the Natural World

So where are those adaptations that we can directly attribute to our understanding and involvement in natural selection and evolution? We need not look any further than the dog. Modern dogs are directly related to one wild variety wolf. But over the course of the last several thousand years, these vicious creatures have become "man's best friend." Why?

The original wolf scavenged and hunted for food. In some cases they did this in close proximity to early human settlements.[49] [50] Through managed-selection these animals were bred to be calmer and friendlier. What dogs lost in terms of their natural aggression and hunting skill they gained in a steady diet and a passivity to live longer and breed more generations of calmer offspring. In return, humans gained an ally that could assist shepherds, alert them when danger approached, and even babysit.

Later on as human civilization grew and our interest in breeding for long or short-legged, black, white or brown, or small or long muzzled dogs became apparent, we adapted the species to fit our needs for either functional utility or subjective whimsy. We now have more than 150 recognized breeds. But all dogs are directly and genetically related to one wild wolf species.

Let's also look at wheat, barley, and rice. These staples of the modern human diet have their antecedents in the fertile alluvial plains of the Middle and Near East beginning about twelve thousand years ago. Although there is evidence for subsistence farming thousands of years earlier, true settlement farming occurred only when humans began to lay down roots and could stay put for longer periods of time.

Early crops were difficult to maintain. There were no Farmers' Almanacs or weather satellites to help guide the first farmers. This meant hit or miss crop yields that resulted in feast or famine. But somewhere along the line some unknown but astute folks started breeding plants and animals for their own use. They knew nothing about Mendelian genetics or Darwinian evolution, but they began to pair their crops and interbreed strains for varieties that were healthier or hardier or that produced greater crop yields, all of which were useful in feeding the growing sedentary populations.

Map of Early Farming & Algerian Cave Painting Showing the Work of Early Dogs Working with Humans in the Hunt

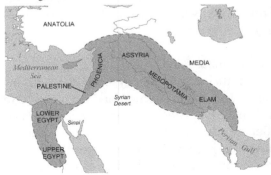

Once humans began to produce larger, more abundant and resilient crops, it wasn't long before we went from tribal encampments to city states. That's because our forebears harnessed artificial selection to benefit larger and larger communities. The adaptation of overall stronger crops was the equivalent of moving civilization from first to fourth gear, going from zero to sixty in about four seconds based on the geologic time scale.

In our consumer society, the term "evolution of products" could be easily confused with biological processes of evolution. Certainly we've come very far from the payphone to the iPhone, or the phonograph to the MP3 player, or even the hearth to warm our homes to modern HVAC systems. Yes, these products have changed over time, but not from a biological perspective. They've changed because of consumer pressure for more advanced technology to bring comfort to our home, work, and social lives. Consumer products have been made better due to planning by human engineers, and sometimes even the complete scrapping of one type of system in favor of building of one that is entirely different. That is what happened with changes in recording technologies. Analog recordings of sound done on pressed vinyl were scrapped in favor of brand-new digital recording systems. This was not a case of an old system receiving many tweaks but a wholesale scrapping of one technology in favor of another. This is the opposite of natural selection.

Where humans have made significant changes using genetics is in our food supply. While it is obvious that humans have been changing and adapting crops for thousands of years, modern Genetically Modified Organisms (GMOs) have been on our shelves for decades. In fact, one of the most controversial foods approved for human consumption is known in the anti-GMO community as the "Frankenfish." This salmon was genetically modified and after years of development was finally approved by the United States Food and Drug Administration (FDA) as safe for human consumption.[51]

The controversy over GMOs will continue because of unsubstantiated worries about the long-term impacts of eating animals and plants that have been artificially modified. Likewise, there are those who worry about the possible long-term impacts of GMOs on the environment. The latter concern has some substance to it, not because GMO crops are inherently bad for the environment but because they can be misused. However, the upside of GMOs is that modified foods could feed a larger population, alleviate hunger, and even reduce our dependence on pesticides and fertilizers, depending upon which genetically modified organisms are farmed.

Many of our modern medicines have been developed, modified, enhanced, or invented using the biology and chemistry of genetics. With the exception of home or natural remedies, the pills we swallow, liquids we ingest, or salves we apply all have been created using the methods of science, including evolution. This includes the bonding or modification of natural and non-naturally occurring genes or chemical elements to create new or advanced materials and medicines.

Genes have been spliced to create fluorescent pigs that aid in the research of human organ transplants.[52] And in Japan the fluorescent gene has been used in cats to study and find medical uses to fight HIV and AIDS. These same fluorescent genes are also being used in marmosets to find cures for ALS, Parkinson's disease, and Huntington's disease. In 2009 the FDA approved a genetically enhanced drug synthesized from goat's milk to alleviate blood clots during and after surgery as well as in childbirth, where blood loss and blood clots can cause imminent death to the patient.[53]

Some Final Thoughts

We are each the product of evolution and we have evolved as a species to be able to manipulate the conditions of life on the planet. We owe it to ourselves and to the creatures who share our home to not be reckless with the earth. We are not separately created, and we share our lot with the rest of nature. Even Darwin realized this as a matter of observation and research. He eloquently noted, "We must, however, acknowledge, that man with all his noble qualities still bears in his bodily frame the indelible stamp of his lowly origin."[54]

Darwin's work cautiously reminds us not to grow too big for our britches. If we do, our species will surely pay the price with our own extinction. Even minor fluctuations in this gingerly balanced world can have small or devastating effects on the conservation of species and the sustainability of the planet. Knowing about evolution means that we have an ethical responsibility to carefully work with it and not damage ourselves or harm others in the process of maintaining our civilization.

It was the brilliant and sorely missed Carl Sagan who said that we have (randomly) evolved so that the cosmos can know itself.[55] This is a startling tidbit of consciousness-raising and self-awareness that takes up little space on a page but speaks volumes about the moral authority of humans and how we are ultimately the caretakers of our destiny and the destiny of countless other species on our planet.

Knowing about Darwin and the people who shared his excitement for natural selection helps us understand the history of the theory but also connects us to the human story of discovery that has been at the forefront of human endeavor. Since we left the trees, banded together, created language and culture, and became the dominant species on the planet we have been adding knowledge to the lexicon of human understanding.

The end product of this long yet ongoing process is what archaeologists call our "modern civilization." Our evolution took millions of years to occur and it is ongoing. It took less than twelve thousand years to create MTV and the atomic bomb. Our species has survived at least one near extinction caused by the geological forces of nature that have humbled many societies and species in the past and will humble us in the future.

So our journey for the rest of the book begins in the next chapter, a review which discusses our common history going back just a few hundred years. This was a time when slavery still existed in America and was based in large part on biblical ideas of the fitness of the races, and when European imperialism left its flags, missionaries, and armies on continents in South and Central America, Asia, and Africa. A synthesis of the natural sciences seen as a candle burning in the dark was about to explode like a powder keg.

It is our hope that you will enjoy the rest of the book as much as we have taken the time to research, cite, and retell Darwin's story and that of his apostles.

CHAPTER 4

LIFE AND TIMES OF VICTORIAN SOCIETY

"History, despite its wrenching pain, cannot be unlived,
but if faced with courage, need not be lived again."

Maya Angelou (1928–2014), Poet

Introduction

Charles Darwin and his apostles lived in Victorian society in England and the United States. Their work and lives can be best understood within the context of Victorian culture and the rapid changes that were taking place. It can be said that every new century is a bridge to the past, present, and future. The nineteenth century was founded on old assumptions and suppositions and ended on so many new realities.

By the 1890's the idea that social classes were and should always be fixed was slowly on the wane. Also on its way out was the age of *laissez faire* politics and economics, especially related to what is described as the social safety net. By the end of the century, new and more progressive politics were taking hold and shaping education, providing social assistance to the needy, and ensuring that workers and especially child laborers had protections to maintain their safety on the job.

The Industrial Revolution would make Great Britain an economic powerhouse throughout the nineteenth century. The British military would fight wars and remain a force to be reckoned with globally, even though, as the century came to a close, nations around the world that had been protectorates of England would seek both legal and political emancipation from the British Empire.

One the most interesting cultural aspects of British society during the nineteenth century has to be the fascination with and the acceptance of both the occult and the supernatural. Across British society people were captivated by "otherworldliness" and the mystical. This began at the very top of the British social class and trickled down to the rest of society.[56]

This chapter is a quick exposition of one hundred years of that nation's social, economic, and political change, which we hope will captivate as well as provide a basic understanding of what life was like for the masses, for the wealthy, and for Darwin's social class. We thus open the door on a century now long gone but still strangely familiar in terms of human wants, power, and patronage.

The Victorian era in both Great Britain and the United States was a time when industrial and technological fortunes were made and lost. The economy rose and fell often, taking peoples' lives and fortunes with it. It was also a time when science was rapidly developing and changing everything that people thought they knew.

This scientific progress took place even though at the beginning of the era, the scientific establishment was largely made up of wealthy amateurs and clergy who did some nature work as a hobby. By the end of the era, due in part to the good work of Darwin and the apostles, science was considered to be a serious, paying, independent profession.

It was a time when religion was changing, and blasphemy laws in England were only recently changed though not abolished or repealed; and when the British and American freethought movements were growing and challenging the power brokers of their respective societies.

It was also a time when collecting was a mania shared by amateurs, universities, private collectors, and even governments. Both Darwin and Wallace collected beetles, and Wallace attributed both his and Darwin's discovery of evolution in part to their beetle collecting. The Royal Botanic Gardens at Kew, run first by Joseph Hooker's father and later by Hooker himself, ranked as the ultimate sort of collection, with the monetary resources of a national government and the international reach of an imperial power. The garden is a tribute both to science and to Victorian triumphalism. [57]

It is no wonder therefore that Richard Owen, the biologist and the creator and curator of the natural history collection at the British Museum, and therefore a rival collector on a grand scale, tried in various ways to get the botanical collection at Kew under his curatorship.

The British middle class plays a large role in this narrative. They were not landed and titled aristocrats nor were they laborers. Those in this class were expected to *become* something—not just run their estates (or their countries) as aristocrats did, or work and hope to get by as manual laborers did. These were people who saw bright futures for themselves, but only if they were educated to the best of their abilities, strived to the best of their abilities, and got lucky. They had the opportunities to be educated and to strive that many others did not. They were also familiar enough with how to become educated that they were all remarkable in their self-education. This was part of what made them successful as scientists. These were also people who understood power and influence and who used it as best they could to promote the best idea in the world. Darwin and all the apostles were members of this middle class.

For the sake of our book, however, this is a story about scientists and the triumph of scientific observation and investigation, which upended a world order that sought refuge in superstition and the teaching of religious doctrine as fact. It is the story of how brilliant men and women of science became real show-me-the-data kinds of scientists who helped support the acceptance of evolution by natural selection. In time, the scientific world came to understand that Darwin and Wallace's work, as well as the work of the rest of the apostles, is the key to understanding the mechanics of nature: from speciation, adaptation, and extinction to allopatric and sympatric evolution, and so much more.

It is the story of the dedicated people who came to understand that evolution by natural selection is the *only* way to fully understand biology—the only thing that makes it make sense. And because it made so much sense, Darwin's fundamental ideas about the nature of nature have led to our modern world today in not only biology but social justice and perhaps even secular democracy.

A Shift toward Mysticism

During the nineteenth century, while many European nations remained sure footed and well established in support of reason, the United Kingdom under Queen Victoria was changing dramatically in quite another direction. For generations after the emergence from the Middle Ages, British rulers were certainly influenced by the rationalist and empiricist schools of thought and by such philosophers as René Descartes, Immanuel Kant, Francis Bacon, John Locke, and Baruch Spinoza. England was essentially managed and governed by rationalists from the mid-sixteenth until the early nineteenth centuries.

However, during Victoria's reign there was a distinct move toward ideas and ideals founded on the supernatural, in part because of romanticism. There were many sincere efforts made within British society to validate mysticism as real. These initiatives, found in the literature of the time as well as in common beliefs, were not limited to select quarters of the British upper class, and it was not unusual for people in many walks of life to seek out the advice of mystics, clairvoyants, and other associated charlatans.[58]

The shift away from reason had several causes. First was genera-
tional rejection of the "old ways of thinking." If we consider how teen-
agers look upon their parents' matured values, we can understand the
inter-generational shift that took place during Victoria's time as queen.
In addition, the world was changing dramatically in Western Europe
and the United States. Industrialization was advancing rapidly and
there were challenges to both government leadership and to the rigid
forms of classism that stymied the lives of millions. These in turn led
to dramatic shifts in social power and to social disruption. So in a way,
relying on magical thinking was (and remains) a great way to buffer
oneself from the realities occurring outside one's own window.

Secondly, Queen Victoria herself was both deeply interested in and
accepting of the occult and mysticism. An avid believer in strange stories
and otherworldliness her whole life, some consider her obsession with
mystics as strengthening both prior to and then growing more intensely after the death of her husband Prince Albert. Rarely did a week pass
when the queen was not entertaining the occult or visiting mediums to
contact Albert or to divine the future for her family and her nation.[59]

While this may seem strange for a sophisticated, powerful woman
to be enamored by the occult, we have to remember that the pull toward magic need not be confined to any social class, national boundary, or education level. In more recent times during the U.S. presidency of Ronald Reagan, First Lady Nancy Reagan had a house astrologer
named Joan Ceciel Quigley. Ms. Quigley was called many times to the
White House for advice, especially after the attempted assassination of
Reagan in 1981.[60]

But during Queen Victoria's time, Great Britain was consumed by
séances and with reading the lines on one's hand to tell the future and
the bumps on one's head to assess intellectual worth—as well as with
tea leaf readings and all forms of magical practices. This movement
also spoke to growing discontentment and disillusionment with the
doctrines of the Church of England. As people mentally rejected the
dogma and literal interpretation of the Bible, they still wished for and
sought supernatural guidance. Mysticism offered an alternative form
spiritual thinking and belonging.

In fact, non-science was alive in England as early as 1820, almost
two decades before Victoria's assumption of the British Crown. For example, phrenology was a pseudoscience founded on subjectivity and
racial bias. Phrenology is essentially the "reading" and comparison of
bumps on the human skull. This was believed to "scientifically" deter-

mine one's intelligence and worth.[61]

Phrenology was taken so seriously that the results could define or even redefine how you were perceived by friends, family, and business acquaintances. It was also used to support racist views and economic imperialism. Looking back at it now, it should come as no surprise that the "experts" in the field found few if any of the right shaped bumps on the skulls of Africans, Middle Easterners, East Asians, and the indigenous peoples of Central and South America.

It was not just the aristocracy that embraced the supernatural and the quack science of the time. Since the acceptance of such ideas was overtly encouraged, they trickled down easily from the British monarchy into the upper and lower classes of British society. It made sense at the time that if you wanted to seem sophisticated, then following the queen and the aristocracy down the supernatural rabbit hole was an easy path.

In fact, choosing the occult over reason could even make you more upwardly mobile and respected within the highest echelons of British aristocracy and government. Think of this acceptance as an updated Pascal's Wager of sorts. Even if you did not wholeheartedly believe in mystics or the occult, going through the motions could lead to access to British royalty, to those who resided in the halls of power, and even, peculiarly, to the leadership of the church.

While the competing slant toward occultism was taking place all across England, Protestantism provided tacit approval by adapting to this age of unreason, creating its own spiritual societies. Religion has a social Darwinist integrity when it comes to adaptation. To avoid becoming extinct, it will change to fit the mores and values of the day. In response to social change and social pressure, organized faith has bent itself throughout the ages, and this has resulted in countless reinventions of both dogma and religious doctrine.

In response to the growing interest in spirituality based on magic, two fundamental shifts took hold among Protestants. One was the start and growth of Evangelicalism, a sort of doubling down on the supernatural side of religion and spiritual idealism.[62] The other was the acceptance of magic-based spirituality with the approval of Anglican clergy. Anglican clergy founded or supported theological initiatives such as The Ghost Society and various Hermetic Societies, as well as The Society for Psychical Research. In fact, this last society worked to investigate and also link clairvoyance, apparitions, ESP, and a plethora of spiritual phenomena to both science and faith.[63]

These shifts in cultural ideals and beliefs regarding the occult could not have occurred at a worse time for science and those scientists who would bring dramatic intellectual change to the end of nineteenth century. This type of explosive acceptance of unseen spiritual worlds or supernatural explanations for nature usually comes with a price and will always have its debunkers as well as its admirers.

Darwin's ideas certainly challenged the values of that era. They dismissed the occult and placed a heavy burden on organized faith. Scientific societies and the government itself were challenged to come up with better and more rational ways to describe humanity's very existence. Darwin knew that his groundbreaking ideas would leave him an outcast in many of Britain's most affluent and influential circles. This certainly played a part in his reluctance to speak up and be heard. It certainly stopped him from personally debating his critics once *Origin* was published in 1859. It also led the book's critics in established scientific circles who were also part of religious and spiritual communities to seek alliances not only with each other but with the monarchy as well.

Britain's Military and Global Power

"The sun never sets on the British empire." This truism has been used to point out that for a duration of many decades, the political and military holdings of the British Empire spanned the globe, so that there was always a part of it where it was daytime and the sun was up.[64][65] The wealth and power of Great Britain did indeed touch every corner of the world. Such imperialism and outright ownership of nations, land, people, and natural resources made it the preeminent world power of the nineteenth century.

At its most influential the British Empire had more than four hundred million people and about ten million square miles of land owned through British investment, corporate raiding, or outright land grabs through military force.[66] Those colonized became subjects of the crown—most became citizens with equal or lesser legal rights compared to those British citizens. Once lands and nations were acquired they were considered to be imperial colonies and protectorates. They served as territories that mother England could use to plunder, barter, incorporate, trade, or fight to maintain.

If you're a fan of the board game or online version of *Risk*, think of your competitor as Great Britain. If this was the nineteenth century, your opponent owns this much of the board: the continent of Australia, New Zealand, India and parts of China, Singapore and Burma,

South Africa, Egypt and Sudan, Nigeria, Zimbabwe, Zambia, Kenya, and Malawi; in the West Indies she owns Barbados, Belize, Jamaica, and St. Croix; in North America she owns all of Canada. And as a reminder, had Britain been victorious in the War of 1812, those American rebels who had won their emancipation in 1776 would once again have been placed under British rule.

The early nineteenth century saw two important conflicts fought by England. The most important of course were the twelve-year Napoleonic Wars that lasted until 1815. The second early century war was one with Russia that ended in 1812. The middle years of the century, with the exception of the two-year Crimean War in the Balkans, were relatively quiet ones from a military point of view. However, beginning in the late nineteenth and certainly carrying over into the twentieth century, nations governed by the British began to demand and fight for their independence.

The waning of the second empire began in the 1860s when Canada won its partial independence in 1867. Australia also became free of direct British control in 1901, and dozens of other nations followed suit in the twentieth century. These included India, Pakistan, and a host of South and Central American, African, Caribbean, and Asian nations.

In total from the late nineteenth century up through today, fifty-seven nations have declared their independence from England. In the game of Risk, losing such valuable global real estate usually means you're doomed to lose the game as well. For the British aristocracy and those sitting at the upper echelons of power in the early and mid-nineteenth century, such losses of territory would have been obscene. Britain thought itself to be the most civilized nation in the world. In fact it saw itself as such a model nation that it seemed right and proper—for the good of the world—to use Britain's military might for outright war and the stifling of colonial dissent.

In some instances Darwin and his grand idea played a role in the shrinkage of British territorial integrity. And in several ways both Darwin's accusers and his freethought advocates may have been partly right. If we accept the idea that humans all evolved from a common ancestor, then this means the Bushman, the Hindu, and those under British rule with dark skin had the same deep ancestors as their British occupiers, no matter how proud the British were of their own pedigrees. It also meant that indigenous peoples in those colonies should be afforded the same human and legal rights as their European rulers.

Natural selection is a biological process supported by scientific evidence. Not only does it refute racism and eugenics, it commands us to understand that no humans can consider themselves to be above and beyond all other humans, based simply on race or ethnicity—even though many have done so in the past and some continue to do so today.

Darwin's theories in biology therefore offer foundational concepts for many constituents, constitutions, and humanist manifestos that liberate and foster equality and democracy. Granted, social Darwinists will make the claim that natural selection only means that the most aggressive and dominant societies can survive. But this is false as shown in the works of Steven Pinker and Paul Bloom. Human beings have lived in packs since before they were human, and societies that reduce intertribal aggression become wealthier, more successful societies. Selection for cooperation generally allows humans (and other species) to become less physically and culturally aggressive.[67] [68]

While this may seem like a very unscientific analogy, the film *Monsters, Inc.* also proves the point in a way that children and adults can agree. If all you're doing is trying scare people to capture their screams, you're probably only working at 50 percent capacity for the amount of energy you're putting out to torture others. However, make people happy and allow them to live without fear and you will get much more cooperation for half the energy.

For evidence, we do not have to look any further than the nations of Northern Europe to see how societies and their respective cultures can live more harmoniously when social pressure is turned down as acceptance and the social network is ramped up to help those in need. Indeed, these same nations have smaller armies and overall less violence than nations that spend billions on armaments and seem almost gleefully ready to invade other nations. In fact, Finland was named the happiest nation on Earth in 2018.[69]

Religious Power

As many historians will point out, the Anglican Church became the supreme church both for king and country as well as for the God worshiping British faithful in the mid-sixteenth century. The rift with the pope had begun during Henry VIII's reign. Henry so despised the Catholic Church that he was outwardly hostile and openly questioned the pope and the Roman church's value and authority, particularly when he wanted to marry Anne Boleyn. So much so that in 1534 Henry was anointed the Supreme Head of the Church of England through the Supremacy Act.[70]

Queen Elizabeth I sought to bolster the Anglican Church split from the Vatican's influence. This us-versus-them sentiment can best be understood by a quote from Elizabeth herself: "What binds us together is not common administration but shared tradition and shared belief."[71] The idea enforced here is that by cultural tradition and religious belief, if you prayed and were Anglican you were an insider and thus were following the right faith. However, if you were a Catholic or other denomination of Christian then you were certainly an outsider, and thereby not only practicing the wrong faith but lower class as well.

Anglican Church teachings have always been serious business. The merger of faith and monarchy meant that following the religion of the empire could bring economic and social prosperity to the faithful. Prior to the nineteenth century, however, at its most extreme, being outside the Anglican Church and its teachings could get one hanged, burnt at the stake, or otherwise humiliated.

As history has shown, British subjects at home or living around the world were sometimes placed on trial for the offense of witchcraft. Many of the accused were murdered in horrible and torturous ways. Today, torture and executions of this type would be seen as evidence of barbarism—the type of degraded behavior that would be perpetrated by ISIS rather than by civilized nations and churches.

The witch trials were and remain violent crimes against humanity. At the trials, "spectral evidence" was admissible in the court. This meant that the accuser had only to see the demons or supernatural forces within the accused "witch" as a vision or dream in order to place a person on trial for witchcraft. It is estimated that in the two hundred years of British witch hunts more than five thousand witch trials took place and between 1,500 and 2,000 people were killed for the crime of witchcraft—all with the support and guidance of the Anglican Church.[72, 73]

Jews did not fare much better under British law as both church and crown fought to maintain the ban on Jewish immigration into the country.[74] The original Jewish expulsion from Britain occurred in 1290 by way of a royal decree from King Edward I. The reason for the action dealt mainly with Jews as moneylenders and their charging of interest on debt owed to their financial institutions. For almost five hundred years Jews were *persona non grata* in England.

In the mid-sixteenth century, with hopes of lifting the ban a possibility, rabbis petitioned the British government, formally requesting Jews be let back into the country. The government, under watchful eye of both the monarchy and church, refused the rabbis' petition. However, while Jews did not have sanction to return to England, many did silently return over the next several hundred years. As they did, they integrated into the social, political, and economic life of the country.

In the nineteenth century the Anglican Church came under tremendous pressure from competing ideas related to human spirituality and social well-being. Aside from the rise in interest of mysticism and the occult, both Evangelicalism and Catholicism were reviving across England and its territories. For these reasons, the social reforms that would become a hallmark of positive change for the masses across Europe and the United States were simply not on the theological or political agenda of the Anglican Church as an institution. Nonetheless, individual Anglican reformers did strive for the abolition of slavery during that time.

It was actually British Evangelicals who called for social reform.[75] The Anglican Church itself was slow to endorse any change related to child welfare, the ending of slavery, or the support of progressive public health and public education legislation. This lack of support by the church held firm, as there was little internal or external motivation to change. The nation's ongoing commitment to a policy of *laissez faire*, combined with a monarch and national government who were vested in the status quo, allowed the church to maintain its views and practices. And it rarely advocated or pushed for progressive social policies during the nineteenth century.

Then as today, there remain two main political contributors to the economic fortunes of the Anglican Church. They include the British Royal family and formal government through the Parliament. Such interdependence requires each member of this well-established troika—the government, the monarchy, or the church—to remain loyal partners no matter the popularity or controversy of any issue or cause facing each of the institutions individually. Indeed, according to the church's own 2018 Cathedral Working Group's accounting, it has received more than 850 million pounds since 1999 through tax rebates from the government. The church elders who are involved with the finance of the religious institution have approximately 8.3 billion pounds under assets.[76]

Money does equal power and in this way both the Anglican Church and the Vatican have substantial wealth with which to help those in need spiritually and socially. From the church's point of view, when organized faith is heavily entwined with the state, and compensated by the state for its efforts, it enjoys a special freedom to do as it wishes with few checks and balances. With such a long track record of mutual cooperation, it is easy to see how the church manages to be heavily involved in English politics and power.

While we will certainly cover the church's response to the publication of *Origin* in greater detail in Chapter 12, it would be impossible to avoid touching briefly on the topic here. As can be anticipated, the details laid down by Darwin confronted not only biblical dogma but those who held that god was first cause even if nature operated by its own mechanisms. The implication of Darwin's work challenged both the religious and scientific orthodoxy of the time. Members of each of these groups had a stake in dismissing natural selection, or pointing out its supposed flaws, to ensure their ideas and social power remained beyond reproach.

As the Vatican would later apologize to Galileo for his permanent house arrest, a sure form of imprisonment,[77] the Church of England eventually apologized to Darwin for the ignorance and defamation of his character and main scientific theories. In both cases however, the apologies came centuries after the emotional pain clouded the careers of these two great men of science. But each church grudgingly came to accept both scientists' grand concepts, an enlightenment which could have come earlier had religious objections been less volatile and ignorant.

We should acknowledge that official Anglican and Roman Catholic Church policy now allows congregants to think and believe what they wish with regard to the creation of the universe and the evolution of humankind. On the other hand, neither the Anglican nor the Roman Catholic Church view natural science without God as first cause of everything in existence, although the Anglicans do reject Roman deification and supernaturalism. In all Western faith traditions and in all church teachings, God is the first cause for the entire universe, the creation of the earth, and the father of all humans. To refute this from a naturalistic and Darwinian perspective is easy, but old religious traditions, like old money, tend to stay within the family.

Social Thought and Policy

Historians agree that during the nineteenth century Great Britain had embraced a form of pronounced philosophical and physical individualism.[78][79] One's social and economic independence from the state was seen as the ultimate form of propriety and prosperity. This was a highly structured and classed society. If one was dependent on the state for support it was seen as individual failure. However, there were few opportunities for individuals to lift themselves out of abject poverty. It was only with the slim chance of entering the priesthood or learning a trade that the poor could hope to change their economic status and, even less likely, their social class.

This zeitgeist of deep conservatism, a precursor to those advocates who would accept Social Darwinism, meant that elites had little interest or inspiration for social and economic change. For the ruling class in the nineteenth century, life was about maintaining the status quo while at the same time doing little for the masses. At least amongst the British affluent class, specifically the monarchy, the parliament, and the Anglican Church, sharing the wealth and social progressivism were on few people's to do list. In fact, the idea of a God-appointed upper class allowed aristocrats to live privileged lives without further justification.

The conservative media of the time, namely the *Quarterly Review* and *Blackwoods Edinburgh Magazine*, steadfastly published stories which upheld the most stringent conservative beliefs, while at the same time each blasted both social reform and the reformers. This meant the conservative view of all things of importance such as politics, religion, social policies, and foreign affairs would all be decided by the reactionary conservatives, much like it has been in modern U.S. politics since 2017 under Donald Trump with the Republicans for a time holding both houses of Congress, the Supreme Court, and the presidency.

But these conservative Britons were not policy wonks so much as true ideologues who wholeheartedly accepted the doctrines of John Stuart Mill and Thomas Malthus amongst other social philosophers. People of power truly accepted that independence from the state was the highest moral good and was best for society.

For the Anglican clergy, the pulp publication of choice was the *Orthodox Churchmen's Magazine,* which was hyper-critical and deeply xenophobic when it came to progressive social change, other religious faiths, and ethnic groups. However, other publications diverged sharply from these views. This divergent media was similar to what we can see in U.S. politics today, with several newspapers and cable news channels taking diametrically opposing views in the culture wars. Then as today, each media outlet served as the conscience of the conservative right or liberal left's views concerning politics, social policy, and social change.

If you were in poverty in England during the nineteenth century, your family had limited, desperate options. The main agencies to go to for help included one's local parish, which was mandated by law to gather the poor and place them into poorhouses. Living conditions in poorhouses leveled off at utter squalor and went down from there. These were the worst slums in the country, where crime was rampant and sanitary conditions were so awful that contagious disease and illness could be found in every dwelling.

The other main alternative for the poor was to go to workhouses. These were pure houses of horror and were deliberately designed to be so.[80][81] Both the elites and social progressives of the time believed that if an indigent person found that his or her only option for support was in workhouses, then that indigent person must truly be desperate. Because of the terrible working and living conditions, workhouses were more like forced labor camps than places that helped people up from or out of poverty.

Poor married couples were broken up by order of the government and sent to different workhouses, the idea being that fraternization would lead to complications and violence. This shattering of the family unit almost guaranteed inter-generational poverty and crime. In our present day, members of poor families that have been broken usually have less access to good medical care, educational options, and social networks that can help alleviate misery and assist people in times of socio-economic stress. So too was this the case for the poor and unemployed in nineteenth century England. Perhaps the only other safety valve available for alleviating poverty was leaving England altogether and starting a new life in the English colonies.

However, if your family had means such as land, money, or access to capital you generally found it much easier to live in the British Empire of the nineteenth century. Then as today, savings meant you were residing in another social class. Because England was so stratified, those with wealth or power typically came from a few groups with established ties to one another through family connection or commerce. These people were the *One Percent* of nineteenth century Britain and included some land owners—among them some who owned vast aristocratic estates—industrialists, bankers, some politicians, and the monarchy. Professionals, including physicians, attorneys, and the clergy were rarely faced with starvation but they were not necessarily wealthy.

If you lived outside of London in the countryside and had means as Darwin and his family did, then life could be very comfortable. Darwin married into the Wedgewood family, who had money through the china and porcelain industry. Thus, even though he came from a family with assets and connections of its own, being a part of the Wedgewood clan meant a life of even greater social and economic security.

If you're a fan of the television program *Downton Abbey*, then you will have a sense of the life that Darwin and those like him lived. Darwin was not a titled aristocrat, and indeed the middle classness of his family had led to his being prompted to go to university and learn a profession. Even so, his was a life filled with servants and caretakers, a life where your needs were taken care of by others. It was a life in which you were expected to hire skilled and unskilled labor to help you throughout your day.

Children's Rights

It would be difficult not to touch on the treatment of children during the reign of Queen Victoria. Children had few civil rights of their own. In fact, with Great Britain in the boom years of its industrialization in the nineteenth century, children were viewed as an abundant labor force that could be used in factories, on farms, and in mines almost endlessly. And why not? With the number of poor families with children to offer exceeding the positions available, child workers were seen as easy to acquire, maintain, and replace by industrial interests in all sectors of the growing English economy.

This "commodity" worked in deplorable and dangerous conditions. Industrial accidents were plentiful, factory air was heavy with waste and fumes, daily abuse from employers was constant, and the work itself was dirty and often unforgiving. Reform only started to come in 1833 with *The Factory Act*.[82] This legislation meant that employers could not hire children younger than nine. It also put into law that children aged nine to thirteen could not work more than nine hours a day. Finally, the act mandated that children must receive two hours of schooling each day.

In 1842 a similar act to the Factory Act was passed by Parliament. It was called the Mines Act[83] and, like its predecessor, it attempted to protect the most vulnerable. For instance, no child under ten could be used in the mines. Before the passing of the Mines Act, it would not be unusual to see four and five year old children working underground. Women, too, were excluded from working in the mines, although the act said nothing about working conditions or inspections. These types of additional workplace protections would not come until much later in the century.

While we can endlessly debate the values of parents who released their children into the custody of factory, farm, and mine, the main reason for their choice was abundantly clear. Children were expected to work and pay for the family's expenses, and many families couldn't afford to do otherwise. With little or no social safety net in place from the government, children had few choices but to become the underage bread winners for their families. The phrase "grist for the mill," which was from the fifteenth century and emphasized the devalued worker, took on a whole new sinister meaning because of the abuse of child laborers in the nineteenth century.

But if children within the labor force had it bad, no child had it worse than those who were wards of state or church. Orphans were seen as a total burden to society and they too lived in deplorable and desperate conditions.

The literature of the time, in particular Charles Dickens's *Oliver Twist*, truly captures for the reader what life was like for a British orphan in the nineteenth century. Orphanages were overcrowded, their sanitary conditions were near non-existent, and food was limited. A life of abuse from other orphans and the workers who ran the institutions was plentiful. Ultimately, the orphanages served to bring several generations of British youth into lives of crime, a pipeline reinforcing social stigmas and stereotypes, which are in some cases still with us today.

If a child was lucky enough to be adopted, she or he had to hope their new families were of the same social class that they were. If they were, then their lot in life and hopes for familial integration were possible. But a child adopted into the upper classes—even if they were from a poor branch of the same family—could expect to spend their lives as second-class family members, often made to be pseudo-servants or at least domestic helpers for their wealthy kin.

Orphans could also hope to be placed in a boarding school run either by the state or the Anglican Church.[84] If a child did get into one of the schools then they would be guaranteed an education. However, the poor living conditions were often equal to that of the orphanages, and of course they provided no emotional support. The latter was seen as unnecessary and unwarranted for these throwaway children who were wards of institutions that cared little for their existence except for how it related to receiving funds from the government for their education.

Just as British adults with financial means led lives outside the shadow of poverty and with much promise, the same was true for the children of the well-to-do and the elites. Family connections, resources, and name recognition meant that children had nannies, tutors, and access to the finest primary and secondary educations that would often lead to acceptance into the most exclusive British universities.

This was certainly the case for Darwin's large family. They had access to the wealth of their parents and thus had the kindest home lives and the finest education. All of Darwin's children who survived to adulthood would become leading academics, business people, and scientists in their own right. And all remained part of the prosperous upper-middle class of British society.

British Crime and Punishment

Until 1856 there was no national police force in Great Britain. Prior to the legislative initiative that created institutionalized law enforcement, there were some paid police officers in major cities and around railway construction. However, in other places those in charge of maintaining the peace and keeping crime at bay were elected as unpaid parish constables.[85] If a person committed a crime, both the victim and the perpetrator were left to fend for themselves within the criminal justice system. Until reforms in the mid and late nineteenth century, prosecutions were mainly begun by civilian victims or witnesses to crimes and not administered by the police or prosecutors.

Those convicted of crimes faced three different types of punishments. The first was being placed in local or national penitentiaries to live out their sentence. The second was to be hanged for their crimes. The third was to be transported to the British colonies to work and serve out their sentence. By the time Victoria rose to the throne, a shift in how criminals would face penance was taking place. Public hanging was popular and it often meant that even minor criminals would face the hangman's noose. However, the increase in executions caused a public outcry for other forms of penalty besides death by hanging.

In 1853, Parliament passed the Penal Servitude Act.[86] This stated that, upon conviction, if your sentence was longer than seven and a half years you could choose either transportation to a penal colony, which was essentially expulsion from the country, or hard labor though penal service, as a way to serve out your term. Since working conditions for prisoners choosing hard labor was just a step or two higher than slave labor, this meant criminals would serve their debt to society as a brutalized workforce.

The policy of transporting prisoners out of England meant that the British colonies would become populated with ex-convicts who, upon serving their sentence, frequently stayed in their new homes rather than return to England where life in the lower classes was immensely difficult.

From the late 1780's until 1868 when transportation was abolished, it is estimated that one hundred thousand criminals were sent from Great Britain to the American and Canadian colonies. But North America was not the exclusive home for British prisoners. Over one hundred and sixty-two thousand convicts were shipped off to Australia to serve out their convictions. In still lesser cases those sentenced to transportation were also sent to the British colonies in the West Indies and as far away as India to serve out their sentences.

For those convicted who stayed in England and did not choose hard labor or transportation, most were sent to work out their penance within the prison system. This fate could be just as perilous as that of convicts in forced labor or in the penal colonies.

With the explosion of crime in nineteenth century England, more than ninety prisons were built between the years 1842 and 1877. Less serious offenders would be locked up in "County Gaols," also known as Houses of Correction. These were generally smaller facilities and were often managed directly by the county. More serious and violent offenders would be placed in "Convict Gaols" run centrally from London. For the prisoners in the Convict Gaols, where living conditions were equally deplorable, their time within the facility would be short, as many were scheduled for transportation out of the country.

In the first half of the nineteenth century and prior to the expansion of the building of physical prisons throughout the nation, most British convicts lived in Prison Hulks.[87] Hulks were essentially decommissioned sailing ships that were converted into docked prisons. At night, prisoners were chained to their bunks in an effort to stop them from escaping. During the day most of the prisoners were forced to work on the docks. Because of the unsanitary conditions within the hulks, disease spread quickly and it was likely that before your time was served you would die from exposure, cholera, or some other communicable disease or die from violence from guards or other prisoners.

As the twentieth century approached there were consistent prison reforms, many started in earlier periods of the nineteenth century, which coalesced into a doctrine that focused not just on incarceration but also on the rehabilitation of prisoners. This led to several innovations in British prison design and penal administration.

For instance men, women, and children were no longer housed in the same facilities. Prisoners were more consistently fed with better food, and sanitary conditions were improved. Prisoners were made to wear uniforms rather than dirtier personal clothes. In many gaols the incarcerated were actually given time to read and write or were taught to read and write in an effort to help them enter society once they left prison.

Public Health

Few if any experts in the field of modern psychiatry would consider the nineteenth century to be a high point in the treatment of those with mental disabilities. It's hard to look back on this time in history. One wonders how Western nations could treat those who were mentally ill so abominably. The management of those suffering from mental illness ranged from the bizarre to the downright cruel. Such organized and barbaric treatment, which masqueraded as "science," should send chills down anyone's spine. (Even researching and writing about the subject and time period makes the authors feel a bit nauseous.)

Nineteenth century treatment of the afflicted can reasonably be described as institutionalized torture. The psychiatric institutions and asylums that were built were meant to remove those suffering from mental illness from society. Most of those institutionalized would be locked away in cages or closets and chained to walls.[88] Frequent beatings of those held in the institutions degraded them and led to high mortality rates. Little food was provided to those incarcerated and unless the ill had family members who would visit, patients were frequently left dehydrated and malnourished.

It's hard to know who was sicker, those confined to the asylum or those sadists who ran the institutions. Some of the asylums served as a form of public entertainment. Bethlem Royal Hospital, one of England's state-run mental institutions, also known as "Bedlam," had charged admission to the public so they could view the mentally ill.[89] The idea was for visitors to enjoy the freakish displays of humanity and also serve as a warning as to what could happen if one went mad or suffered from mental illness.

Private asylums were no better than those run by the British government. These private institutions at least stated they existed to heal the mentally ill, but none in their care or custody was ever cured and the treatments offered were still comparable to torture. These alleged "cures" included various forms of bloodletting, cold water immersion, near drowning, scaring patients, and immobilizing patients by tying them to beds or chairs and pouring water on them. The fixation with water torture stemmed from the idea, popularized by American physician Benjamin Rush, that water could remove the toxins in the body that caused the person's mental illness.

Healthcare in the United Kingdom today is science based. Doctors and other medical professionals are educated, and they understand illness and care of the ill in ways that are fundamentally different from the views of their nineteenth century predecessors. In many cases we can thank Charles Darwin for liberating the biological sciences, especially as they relate to medical science, since we now understand how organisms adapt and evolve—including the causes of infectious diseases, such as bacteria and viruses. This allows doctors to prevent and in some cases cure illnesses before they become infectious or lead to mass contamination.

While modern British medicine works best by using the microscopic, biochemical, and genetic evidence to diagnose and understand illness, during the reign of Queen Victoria physicians just did not have the science or knowledge of biochemistry, endocrinology, and the need for sterile surgical techniques that we have today.

In some medical textbooks of the nineteenth century it was suggested that "night air, sedentary habits, anger, wet feet, and bad air" caused smallpox, scarlet fever, and measles. Indeed, cholera, a communicable disease that struck Great Britain fiercely in the nineteenth century, was thought to be caused by cucumbers and melons or high emotions like by passion, fear, and rage.[90]

Two major advances in surgical science occurred in nineteenth century England that allowed physicians to perform longer and more risky surgeries. The first was the midcentury use of chloroform to anaesthetize patients so to keep them unconscious for longer periods of time. The second advance was the increased use of sterile rooms and antiseptic tools to perform surgeries. This meant patients stood a much better chance to survive post-op or at least avoid infection during surgery.

As in the United States at the time, the use of effective medicines was still far off. As a result, many curable conditions ravaged populations. Class also played a role in access to good healthcare. And even if one did have means, without a scientific basis for a cure many people turned to scientifically untested herbal remedies, water cures, hypnotism, and even electro-therapy for dealing with illness. If you had a spiritual bent, as many people did during Queen Victoria's reign, it would not be surprising to find both patients and doctors blaming mental and physical illness on a person's spiritual failings.

Education and the Public Education System

What most of us accept as a primary education would seem foreign to nineteenth century English educators, the monarchy, the church, and common folk alike.

Today, we expect that children will begin their public education at around five years of age and attend classes through their teenage years, emerging as college-ready, technical school-ready, or life-ready approximately by their eighteenth birthday.

When done well, public education today serves to make students well-rounded and productive global citizens. This view of public education was not the case in the Great Britain during the reign of Queen Victoria and it took almost a century of trial and error, as well as many controversies, for mass public education to take root.

Early in the nineteenth century, education in England was defined and offered through private means only. This meant only those institutions with power could or would educate youngsters, and specifically for their own institutional needs. During this time, Sunday schools and schools devoted to trade educated the youngest of Britain's masses, and even this was limited in scope. It should be noted that in England the phrase "public school" actually refers to a private educational institution. The original public schools were for aristocrats and were spoken of as "public" only because they were an alternative to private tutors. (For the rest of this chapter, however, the terms public school and public education will refer to government-sponsored, taxpayer-supported educational institutions.

As with all advances in public education during this time, tweaking was more the rule than full-out reform. Thus entered the innovation of monitoring schools which were not sponsored by the church but used the Bible to teach and enforce biblical moral and ethics. These schools taught hundreds of children at once, widening literacy while at the same time inculcating the masses in Christian faith and beliefs.

Still more modification would come through private entrepreneurship combined with the philosophical ideas which placed a high value on education at the earliest age possible. The first elements of daycare were founded in England near the midpoint of the nineteenth century.[91] Infant schools became very popular and they served children from the age of two onward. Not unlike today, the children in these schools were cared for so that their parents could go off and earn a living in cities, farms, factories, and mines. Unencumbered by the need to provide childcare, the parents of these children helped to expand the

British economy during the growth of Britain's industrial revolution.

The constant drum beat of "blame the poor for being poor" made legislating and obtaining land for mass public education sponsored by the state very unpopular. Many felt that teaching the poor would be counterintuitive and would lead to social unrest. The reasoning of course was that if poor people could read and thus understand their place in Britain's economic life, they'd be upset and perhaps even take it out on those of higher station.

While the educational debate raged on, by midcentury an enlightened Parliament sought and received national funding for the purchase of land to erect public schools. In a series of acts beginning in 1841 and running through 1855, funds were used that created dozens of schools across the country dedicated to educating the poor. This of course was a feat of Herculean importance and a process of social engineering unheard of before.

It was also very successful, and so much so that by the late nineteenth century more than 2.5 million children were receiving some form of standardized public education, almost double the number of children receiving an academic or technical education in the second decade of the nineteenth century.

Of course none of these advances in government sponsored public education impacted the lives of the wealthy and their children. For the young offspring of the upper classes, private education in the sciences, mathematics, philosophy, literature, Latin, biology, anatomy and physiology, and theology were of the highest caliber. It had been this way for generations.

Education for girls was another matter entirely. For the lower classes and the poor, little attention was given to girls except perhaps their learning how to lightly read or write, and even that was inconsistent. For those in Darwin's class and above, education for girls was more about finishing and etiquette, making little ladies ready to attract a suitor to marry. If they accomplished that, then they could focus on preparing a home and providing a strong maternal and cultured influence on family life.

A blossoming of education though public libraries and public lectures also offered British subjects, including working people, means of improving their knowledge. However, these resources were inconsistent in availability and quality. They were also of limited utility for moving someone to another social class.

Economic Systems and Technology

Nineteenth century England was captivated by the idea that the government should not interfere in the social and economic life of the nation. It was a time of industrialization, economic expansion, and a form of free-wheeling capitalism that allowed many to become wealthy while at the same time kept the poor in check in dead-end jobs of heavy labor. This sense of unbridled economic growth was best described as a laissez-faire approach to how nations and markets should operate in the modern industrial age.

If you have taken an introductory economics class you know that one of the main tenets of laissez-faire is the idea that nature and all things human made find an eventual harmonious balance. Taken to an extreme in economics it suggests that when government interferes with economic systems as they try to find balance, the result leads to further disruption, disharmony, and the stifling of liberty. Economic philosophers such as France's François Quensnay, England's Adam Smith, and Englishman John Stuart Mill and Thomas Malthus were all deeply committed to the most liberal aspects of laissez-faire.

Also committed to the ideas of laissez-faire were influential British industrialists Richard Cobden and Richard Wright, who both clearly saw economic advantages in unobstructed and unregulated business.[92] They frequently advocated for unbridled trade, since both they and their companies profited from the governmental hands-off approach to their business dealings.

However, the darker side to laissez-faire, both in business and in society, meant that there was little political goodwill to help those in need. Taken to its extreme, the refusal of the government to become involved in the lives of its citizens meant many were taken advantage of economically and thousands starved, were homeless, and lived in squalor. They were the uneducated masses and were uncared for medically.

In fact the rise of Marxism, particularly around the time of the publication of the *Communist Manifesto* and *Das Kapital*, was a reaction to "capitalism gone wild." At its core Karl Marx and Friedrich Engels saw workers as being terribly abused, and their ideas about the redistribution of wealth and reorienting the means of production toward collectivism all came about as a reaction to the growth of industrialization and the structural inequalities so prominent within the laissez-faire doctrine.

One of the worst instances of British laissez-faire policy was the

British government's lack of assistance during the Irish potato famine that began in 1845 and lasted until 1849.[93] After four years of failed potato crops, it is estimated that as many as 1.8 million people died from lack of food and food assistance. Placed into a modern context, this bungling of aid was the nineteenth century equivalent of the lack of assistance in the United States after Hurricane Katrina wiped out much of New Orleans' 9th Ward and other below sea-level parishes out of existence. Except that it was slow starvation that went on for four years. The lack of British support also would create a migration crisis. Desperate Irish refugees fleeing the Great Potato Famine left their homeland and immigrated to the United States.

Technology Innovation

The refrigerator, the radio, television, the microwave, home computers, the iPhone, and the internet are all twentieth and twenty-first century inventions which upended how we live, how we work, how we communicate, and perhaps how we will survive into the future. These creature comforts would perhaps not exist had the world not gone through a period of rapid industrialization during the nineteenth century. Economic growth and expansion across multiple and new markets and industries coupled with strong sustainability and the manipulation of economic policy in part built our global modernity.

What also powered nineteenth century industrialization was a sense of entrepreneurship, inventiveness, and trade that not only created great wealth but also expanded the economies of Western nations across Europe and the United States. In England, for instance, the invention of the steam engine lead to greatly expanded trade in textiles, agriculture, iron, and a host of manufactured goods that could be sold domestically and internationally. This in turn increased the incomes of those who worked in the factories and massively increased the incomes of those who owned the means of production, such as the factories themselves.

Aiding the industrial expansion were steam machines that served agriculture and which allowed farms to use a host of new technology for all sorts of work. This included new machines for reaping. Newly powered steam machines also did a host of laborious chores such as milking cows, and tractor and plough steam machines tilled and allowed for faster harvesting in the fields.

In symbiotic allegiance, new techniques and the invention of better machines to dig coal also helped Britain's industrial revolution. Of course this also led to child labor and other labor issues. But because of the need for coal to drive industry, the mining labor force expanded by half a million in less than half a decade and coal production increased annually to more than 230 million tons per year.

New ways of smelting steel more quickly and efficiently also gave the British steel industry a competitive advantage. Steel was central to the manufacture of railroads, armaments, for the making tools, and most importantly for ship building. By the end of the nineteenth century nearly five million tons of steel were being produced, and of that number more than a million tons were sold and exported to nations around the globe.

British inventors made great contributions. For instance, Michael Faraday demonstrated that a magnet could produce electricity. Faraday also invented the dynamo, a machine that served as an electrical generator and could produce and send sustained direct current. The generation of electricity meant that factories could replace steam and coal fired-factories with cleaner and safer forms of energy.

Communications technology also became unbound thanks to Sir Charles Wheatstone and Sir William Cooke. The pair invented the first viable telegraphy system in 1837. Sir Rowland Hill reformed the post office and invented the first penny stamp. This invigorated postal service and letter writing throughout England.

There were dozens of other small and large ideas, concepts, and inventions that owe their genesis to the great thinkers and tinkerers of Great Britain at that time. This one book chapter cannot fully give all those inventors their due.

However, considering its impact on our world view, its unification of the entire field of biology, and all of the downstream effects that it has had, we emphatically state that Darwin's and Wallace's contribution of the discovery of evolution by natural selection is, more than any other discovery or invention in the nineteenth century, the one that has led to our current collective modernity.

While some elements of life would seem odd if we traveled back to the 1850s, in a few respects Darwin's time was not unlike our world today. Then as now, we live in an era where access to information is becoming much faster and easier to obtain. Media and journalism move public opinion, and scientific collaboration is usually seen as a virtue and not a detriment. This is the context in which Darwin's apostles lived and worked throughout their lives.

Other Social Policies and Problems

The topics discussed in this chapter offer the reader a quick glimpse into the social, political, religious, and economic lives of those living in Victorian England. It would be impossible to discuss all aspects of a society clearly and cogently in the confines of a short chapter. However, there are other areas of nineteenth century British society which admittedly have gone untouched by this rather global review.

Although we have talked a bit about children, particularly girls, being treated as lesser and collateral beings, we have reluctantly not touched on the start of the women's suffrage movement or the fact that the United Kingdom was in the past a slave-faring nation. We have only lightly touched upon the devout imperialism, classism, and racism found in nineteenth century British culture. Nor have we discussed in detail the ethics of colonialism and imperialism and how the use of military forces, church missionaries, and enhanced economic power were used deliberately to overwhelm and subsume other people's languages, customs, and cultural traditions.

Conclusions

It is said that no person is an island. We can certainly look back on the nineteenth century as a less hospitable and more dangerous time, especially if you and your family were poor or were part of the lower classes.

Should we hold the past and the people who lived in it accountable for their actions as well as their failures? The answer is, yes, sort of. Had we lived in the nineteenth century would we have thought or acted much differently?

This is the conundrum of historical review. We can see clearly the history but have no way to go back in time to change it. Ethically, the question also arises, if we could, should we change it? Looking into the future we can also ask the equally serious question: How will people in one or two hundred years look back on our global culture today? Will we be seen in hindsight as equally barbaric and emergent?

We do know our common humanity survived the nineteenth century with its good and bad, its hostile aspects and its emerging liberalism. Every generation is given the choice to learn from or repeat the mistakes of the past. Given that we have learned so much and have so much documented about the people, times, and era, perhaps its best to see the United Kingdom in the nineteenth century within the context of change, understanding that we cannot in all fairness judge the people of previous eras through the lens of our lives today.

We can be grateful for the times in which we live, for so many liberal secular constitutions, and for the United Nations Declaration of Human Rights. We should be grateful for the third iteration of the Humanist Manifesto, for the medicines we now have, and for the technologies that today connect people and their ideas as never before. If we are gracious and mindful of how far we have come, and determined in our efforts to learn from the past and from current science as well, then we are certain that it is possible to create for ourselves, our children, and for generations to come, a future that will be safer, richer, and kinder.

CHAPTER 5

DARWIN'S POWER FOR REASON AND KNOWLEDGE

"All our knowledge begins with the senses, proceeds then to understanding, and ends with reason. There is nothing higher than reason."

Immanuel Kant (1724–1804), Philosopher

"Faith consists in believing when it is beyond the power of reason to believe."

Voltaire (1694–1778), Philosopher and Historian

Introduction—The Power of One Book to Change the World

There is much already known about Charles Darwin as both man and scientist. However, our purpose here is to offer new facts, details, and insights about one of the most important figures in modernity and modern science. It is part of a proud lineup. In fact, one can argue that there is a cottage industry of biographical, contemplative, reverential, and critical books on Charles Darwin and his superb and all revealing theories and writings.

For instance, a search of the national Online Computer Library Center (OCLC), the bibliographic database used by the Library of Congress to catalog library materials, has over 2,300 items on Charles Darwin. There are more than 24,000 books and other materials on human evolution, and more than 1,400 books on *Origin of Species*.

In total, the bibliographic universe includes about 27,000 items on Darwin. If we included books on modern biology (books after 1890), a research area that owes its very life to the concepts developed by Darwin, we would need to count another 422,000 books into this influential Darwinian collection.

While there has always been a conflict and rivalry between evolution and religion, by comparison if you did bibliographic searches for the New and Old Testament, you'd wind up with about 200,000 items, the subject Jesus Christ about 66,000, and Islam about 275,000 books and other materials. This of course only covers the Abrahamic traditions.

So when people ask, "Why there so many more books about the Western faiths than science," you should tell them they're incorrect in their assumptions. That indeed, because of Charles Darwin, we have come a long way forward and really match, and sometimes even exceed, the number of books written on or about organized religion. However, it speaks to the poor state of science education in this country that some people do not realize that there is *any* evidence for evolution, and most people don't realize how much evidence there is. We have libraries full of evidence for evolution, and most people don't know it.

Why Evolution Detractors Always Lose

Since the time Darwin wrote *Origin* up until today, he has had his proponents and opponents. As we have come to expect, science is a self-correcting process whereby theories are replaced with better concepts and ideas once they are revealed. One reason why even the most ardent detractors of natural selection fail to disprove Darwin's superb idea is because they offer no credible and scientifically valid alternative to biological descent over time as observed and postulated by Darwin.

Indeed, even the most sophisticated theologians and opponents of natural selection can only at their core declare their proof of the falsity of evolution though philosophical and nonscientific means, even when they claim science is on their side. So for those who support concepts like intelligent design, which is the religious notion of an intentional and controlled creation of each species, the mountain to climb to dethrone Darwin and natural selection is so steep that one must inevitably fail, suffering a lack of credibility at every turn.

Why such failure you may ask? The simple truth is that each and every detractor's main argument requires an all-knowing creator to be behind the creation of life, not only on earth but of the universe itself. Such a concept is always theistic in nature and thus, as both reason and the courts have rightly claimed, is un-evidenced and scientifically unprovable.

Evolution's detractors always lose because even after over 150 years of trying, they have still haven't developed a *better* scientific theory of biological origins than evolution. They have completely conceded to science in this respect—that they have never been able to make a successful prediction about the natural world that is based on their theistic approach. Creationism, intelligent design, and all the pseudoscientific excuses for not accepting evolution all fail in this way. They do not make testable predictions. Evolution does.

Darwin's Powerful Insights—The Voyages at Sea and at Home
The Voyage of the Beagle

Charles Darwin's fateful voyage at sea would provide him insights into the operation of how all biological organisms are locked in a struggle to survive, adjust, speciate, and at times become extinct when ecologies change faster than any one species can adapt. This ocean voyage changed Darwin in ways that were profound and indeed would crystalize his methods of observation and research for the rest of his life.

Darwin was just twenty-two years old when he set out on the *HMS Beagle*.[94] His mission was to explore lands fully unknown, collect animal and plant species yet to be chronicled, and document his travels for the consumption of a public eager to learn about the New World. In the process of his travels, Darwin wrote in his journal and adroitly described natural places that were thousands of miles away from England and which were deeply foreign to many of his countrymen.

The journey across the oceans was long and at times difficult. Darwin had to deal with his almost continuous motion sickness. In total, Darwin's five years on the *Beagle* would make him a world-class frequent flyer today. As he chronicled his experiences, his time on the water would amount to three and a half years at sea. The other year and a half were spent on different land-based locations where he wrote down his observations and collected specimens. These animals and plants were preserved for further study back in England.

The *Beagle* visited a diverse group of lands and islands across Central and South America as well as in the Pacific. While most people think of the voyage as a trip that solely included islands in the Atlantic Ocean near the Americas, the fact remains that Darwin and his shipmates also visited New Zealand, Tahiti, and Australia. They also visited the Mauritius Islands near Africa in the Indian Ocean.

While on board, the Victorian naval caste system required that only Darwin be allowed to dine with the captain. They got along well, most likely due to Darwin's aversion to conflict. However, to Captain Fitz-Roy's credit, he is said to have been the person who gave Darwin that first volume of Lyell's book on geology that he read while at sea.[95] This gift can be seen both as a token of acknowledgement and of welcome. The welcome was an important gift, given that the rest of the crew viewed Darwin with contempt, since he was an outsider, an educated man, and a passenger they regarded as an unnecessary luxury.

According to Darwin, FitzRoy did have an aggressive temper[96], but that was mostly directed towards his crew. The few times FitzRoy's ill temper was directed towards Darwin while on the *Beagle* was when they would debate the issue of slavery. Darwin was an abolitionist and FitzRoy, at the time, was not. It became a hot button issue which, to comfort both men, was not discussed much after the few times they debated the topic. The ship was uncomfortable and had tight quarters, so clearly it was best for the sake of survival not to provoke negativity if both men were to survive the five-year journey.

Darwin's keen observations regarding the Galapagos' rich biodiversity were a major part of the original inferences that he drew upon as he developed the theory of natural selection. But during the entire voyage, Darwin took notes and collected specimens. This resulted in his amassing a huge collection of birds, sea creatures, mammals, plant species, rock formations, and fossils. These deepened his (and ultimately our) insights into the biological and natural world.

What became apparent to Darwin as he studied his specimens was the incorrectness of the concept of "the fixity of species." This idea requires all animals to be fixed in their place in nature and says that they are fundamentally immutable and unchangeable. Fixity of species is a scientific holdover from when men of science and men of the church were in charge of the universities and scientific communities that studied the natural world.

The "fixity" concept is based on religious notions of special creation and essentialism. Championed by the father of taxonomy, the eighteenth century naturalist Carl Linnaeus, this bad theistic science idea not only demands all animals be stagnant but so too their environments. This means the ecologies around the globe are also fixed, and there is no geological change, no climate change, and more importantly no such thing as species extinction.

Charles Darwin rightly confirmed that ecological and geographic pressures do cause species to adapt and change over time. This concept, known collectively by the key concepts of genetic flow, genetic drift, and speciation, means all surviving species are the biological byproducts of generations of genetic change and adaptation.

When animals cannot adapt or speciate they will eventually become extinct. This may happen quickly, much like sixty-five million years ago with the six-mile wide asteroid strike that killed off many species, including most dinosaurs. Still other times, like within our own species' evolution, there may be multiple genuses that are occupying the same or similar niches over hundreds of thousands of years, until a few or even one species becomes dominant.

The shedding of old misconceptions about how nature operates remains fundamental to the story of Darwin's development of his theory of natural selection. But there were many other influences that would eventually cast *Origin* into world history as a cornerstone of scientific literature. Darwin's tenacity as a naturalist and observer of nature for one, combined with his steady all-inclusive methodology composed of two decades of research at Down House, helped to build a foundation for his theory.

The world around Darwin was maturing in ways that made it ready for *Origin* to be consumed and digested by both the scientific community and the general public. At a time when political and social upheavals were occurring all over Great Britain, new ideas about the "fixity" of social and economic life were also being challenged.

But in 1836, when Darwin returned to England from his sea adventure, there was no way for the man to know what his observations would mean and how they would change his life and modern science. Darwin may have begun his voyage on the *Beagle* unsure of himself, ill-footed in terms of his identity and his future, but when he returned to England he was stronger and ready to devote his life to describing his numerous collections.

Examples of Darwin's collected finch species found on the Galapagos Islands.

Darwin's Literary Productivity

There is a saying about books. One written by an author is an accident, the second is a coincidence, the third a trend, and by the fourth a legacy. In total, Charles Darwin published fourteen books on his travels with the *Beagle* covering the zoology, the geology, and the fossils of South America prior to 1859. His books also covered the diversity of flora and fauna discovered on his travels around the world. Darwin's many friendly associations also paid off during this process. His numerous collaborations offered many of his colleagues a chance to review, edit, and critique his works prior to their publication.

In addition, Darwin was a letter writer in the grand Victorian tradition. His copious correspondence with friends, family, students, and colleagues has been amassed in several archives and has been well documented over the decades of his life.

Letters to and from each of his many fellow researchers show a deep respect for science and for many of the core concepts of evolutionary theory. The mid-nineteenth century was a time when the pace of discoveries and theories concerning the natural world was exploding across Western Europe and in the United States, especially in the area of biology and the physical sciences. So it's no wonder the ideas championed and developed by Darwin over the next twenty years were so critical to his contemporaries.

As noted earlier, prior to and long after the publication of *Origin*, both Lyell and Owen would break with Darwin and the tenets of natural selection—not only for theological reasons but because of their firm commitment to the earlier concept of the fixity of species as well as their belief in God. Fixity not only implies species do not evolve but it also ensures a "creator" at the center of all creation. This was championed first by William Paley, who in 1802 in his book, *Natural Theology; or, Evidences of the Existence and Attributes of the Deity*, came up with the analogy of God as the grand watchmaker.[97] Often, when faced with a new scientific paradigm, it is not unusual to have really smart members of the science community stand firmly against the changing tide of new discoveries.

One can and should take natural selection as worthy of acceptance today. However, science is full of checks and balances and natural skepticism. As a result, any new theory will have both its early adopters as well as those who reject its new ideas. However, rejecting a theory based on religious faith is clearly an error. For instance, after the publication of Albert Einstein's theory of special relativity, there were

many physicists who remained dedicated to the concept of the elec-
tromagnetic origin of space/time or other non-theological criticism.
There were also scientists who thought that the level of acceleration
proposed by Einstein's theory was physically impossible to obtain or
maintain, thus they considered his mathematics and the law he proposed null and void.

Lyell said that his main objection to evolution was that he felt that it
was just a re-working of Lamarck's discredited idea of transmutation.
This idea held that acquired characteristics could be handed down to
offspring and successive generations. In addition, Lyell was a Christian
who at the time of *Origin's* publication could not square evolution with
his core religious beliefs.

Lyell's objections aside, he did help both Darwin and Alfred Russel
Wallace publish and present their research work in 1858 at the Linnean Society of London, a year before *Origin's* first publication. Ironically,
Lyell was also a dear friend of both T. H. Huxley and Joseph Hooker,
two of Darwin's apostles. But he remained cool to the concept of evolution for years after Darwin published the book that would change biology and toss God out of the process of the development of life on earth.

By the time of *Origin's* publication, Richard Owen had become a
deeply respected comparative anatomist and also the head of the British Museum's natural history collections. Owen truly accepted some
form of supernatural force or divine energy for the creation of life.
Thus, because natural selection removes humanity from what Owen
considered our natural place at the top of the chain of being, Owen
concluded any absence of faith in *Origin* was a mortal blow to the divine. He thought that evolution as described by Darwin was truly a
dangerous theory mainly because it "bestialized" humanity and top-
pled our species from what Owen considered to be our rightful and
divine place on top of all creation.[98]

The Darwins at Home

To be clear and also to avoid any sense of misogyny, while this book
does focus on several men who directly debated and advocated for
Darwin and his theory of natural selection, there were women in his
life who were also critical to Darwin's success and memory. These significant pillars included his wife Emma, who he married upon his return from the *Beagle* adventure and who helped review and edit many
of his earlier works, including *Origin*. Also, his third daughter, Henrietta Darwin, was a keen writer in her own right and would help edit

Charles Darwin's *The Descent of Man,* and two Darwin family biographies. These included *The Life of Erasmus Darwin,* Charles' paternal grandfather, and *The Autobiography of Charles Darwin,* which was published five years after Darwin's death in 1887.

Starting in 1839, Charles and Emma had a total of ten children. All told, three of their children would die before the age of eleven. Their daughter Anne Elizabeth would die of tuberculosis in 1851 when she was ten years old. The Darwins' second daughter, Mary Eleanor, was born in 1842 and lived slightly more than three weeks before passing away. The last child, son Charles Waring, was born in 1856 but survived only about eighteen months.

Possibly it was the death of Anne Elizabeth that fully broke Charles Darwin's heart and pushed him further into the view that there is no divine presence operating to improve the world. From all accounts, Anne was Darwin's favorite child and her loss was just too much for him to absorb. Darwin took Anne for a variety of "leading-edge" nineteenth century medical treatments while she was ill. These included water treatments and bloodletting. None of them prolonged her life, and in fact each may have worsened her condition. In her last week of life, Anne was in terrible pain and she suffered greatly. Her agony, more than his research into the biodiversity of life, caused Darwin to fully lose his faith in God.[99]

By all accounts, Darwin was an attentive father to all his children and a devoted husband to Emma. The Darwin brood would often join their father on scouting expeditions at Down House, the Darwin family home and estate. Evenings were filled with conversations about the specimens collected and about nature in general. Always painfully anxious and reclusive in the company of others, Darwin was right at home leading and teaching his children. And his children learned well and did well later in life, providing they did not come to any illness.

Emma and Charles' first born, William, became a successful banker and businessman. William lived until 1914. As mentioned previously, Henrietta not only published successfully but also edited her mother's letters, which were published in 1904. Henrietta lived until seventy-four. George Darwin studied astronomy and mathematics, became a Fellow of the Royal Society, and also became an attorney. He passed away in 1912.

Their son Francis Darwin would become a well-respected botanist. He also helped his father with his writings. Francis published *The Life and Letters of Charles Darwin* and would die in 1925. Charles and Emma's son Leonard would have the longest life of all the Darwin children, stretching ninety-three years and ending in 1943. Leonard would grow up to be a military officer, politician, and economist. Their last son, Horace, would study engineering and design science instruments, as well as become the mayor of Cambridge. His death came in 1928.

Only three of the Darwin children had children of their own. George would have four children, Francis would have two, and Horace would go on to have three. This means that any person living today who claims direct ancestry to the Darwin-Wedgewood line would be a descendent of one of Darwin's nine grandchildren.

Darwin's Research and Cultural Associations

Charles Darwin's impact on our knowledge of the biological world is so significant that without his work (and Wallace's), our reality could not be the same. From medicine to plant and animal husbandry, from evolution to literature, the events that culminate in our modern way of life stem from how we understand, describe, and communicate what we observe and what science has taught us about the planet and ourselves. Darwin, the keen observer, would be impressed by how his almost 160-year-old book has in large part informed our collective past, present, and future.

His books, letters, and ideas are certainly a huge part of his legacy. But so too are the places now named for Darwin to commemorate his scientific achievements, the places he visited, his research, his observations, and his writings. For instance, the Charles Darwin Research Station in Ecuador works to conserve species of the Galapagos Islands. The city of Darwin in the Northern Territory of Australia is named for the biologist. So too is an area of the Falkland Islands. Another nineteen landmarks around the globe bear Darwin's name.

There are over 250 species of animal that have, within their taxonomic name, reference to Darwin. As of this writing, Charles Darwin remains on the British ten pound note. There is a UNIX operating system named for him, as well as a proposed series of spacecraft. The Royal Society awards to serious researchers the prestigious Darwin Medal, and the ironical Darwin Awards are "given out" to people who accidentally remove themselves from the human breeding population in silly, downright stupid, or in unusual ways.

Darwin's first major published book hit the presses in 1838 and was entitled *The Zoology of the Voyage of H.M.S. Beagle*, which was gigantic in size, effort, and merit. This was no ordinary work, coming together as five parts and nineteen numbers. Each part was written by an expert in a particular sub-field of natural history and was based on Darwin's specimens and findings. Each volume was chiefly edited by Darwin prior to publication and could not be published without his approval.

Part one of *Zoology* focused on "Fossils" and was drafted by Richard Owen. Part 2 was entitled "Mammalia" and was written by British naturalist George Waterhouse. Part 3 focused on "Birds" and was penned by ornithologist John Gould. The fourth part was on "Fish" and was developed for publication by naturalist Leonard Jenyns. The final part, entitled "Reptiles," was completed by English zoologist Thomas Bell. This fifth and last volume of the work was not completed until 1843.

It is also about this time that Darwin began to think deeply about and refute the concept of "transmutation" as a false start in how species adapt and change over time. As early as 1838 it is in his famous field notebook "B" which includes the iconoclastic drawing showing a visual representation of how the concept of evolution would be shaped.

The image scrawled by Darwin shows how the theory of natural selection would eventually be built based on the process of descent through modification.

If 1838 was a banner year for the young naturalist in terms of publication, then the following two years can best be described as literary interest built on his years of research. The year 1838 also saw the publication of "Observations on the Parallel Roads of Glen Roy," a paper Darwin wrote on the geology of Scotland.

In 1839 Darwin wrote the *Journal of Researches* (finally retitled as the friendly *Voyage of the Beagle)* and he was also elected to be a Fellow at the Royal Society. Always a prolific thinker and author, Darwin would publish dozens of articles and books on a wide range of his observations. While London served as a space to lay the foundation for his reputation as a naturalist, it would be at his home at Down House that his observations and writings would make him a legend.

In 1844, Darwin would publish *Geological Observations on the Volcanic Islands Visited During the Voyage of the Beagle*. This book and another each looked at the natural geology of South America, in particular the Falkland Islands, as well as Australia and South Africa. In 1849 Darwin would be asked to write a chapter on "Geology" for the British Navy for use in a scientific manual.

The year 1851 proved to be a very productive year for Darwin's research and his literary career. By now Darwin had been living comfortably at his estate at Down House for almost a decade. This sanctuary offered him the time and privacy to study and write about his meticulous observations. Sadly, it was also the year that his dearest child, Anne, would pass away (in late March). Her death at such a young age sent Darwin into a deep depression from which he would never fully recover.

During these years Darwin's research mainly focused on the natural history and comparative development of barnacles. The first of two volumes from two separate books would be published, *A Monograph on the sub-class Cirripedia* and *A Monograph on the Fossil Lepadidae*. The second volume of both of these books would each be published in 1854.

His copious letter writing and increasing visits from his apostles grew more intense as Darwin began writing *Origin,* which was based on his field notebooks, his observations, elements from all his previous manuscripts and books, and from the vast information and collections from other natural historians and scientists with whom Darwin communicated.

During this time, Darwin became more and more of a recluse. He chose to spend most of his time at home by himself involved in his research or in taking undemanding trips with his family for health reasons. This is what makes Darwin such a puzzling but fascinating subject: a loner with many friends, a scientist willing to spend days avoiding his family for the sake of his research, yet a beloved father who loved his children and wife. A lost soul and wild child early in his life, he would become one of the world's most vital scientists ever to propose an idea about the fundamental operation of the nature of biology.

Darwin refused to publish his ideas about evolution for many reasons, even as his writing on the subject crystalized and was updated over the course of his research. Darwin rightly knew that he would open himself to criticism from many quarters. He understood that his theory would upend his respected career in British social circles and within the intellectual community of his fellow naturalists. The book would cause a final break with the church, a relationship that had become tenuous at best since the death of his daughter Anne. Of course the work would also deeply hurt his supportive and unwavering Christian wife, Emma. Finally, being in the spotlight was never what Darwin wanted and as he aged he became more reclusive but remained very productive till the end of his life.

Natural selection was ready not so much to be "discovered" but ready to be *uncovered* by humanity. Darwin and Wallace, two committed naturalists who were contemporaries who respected science and nature, were each ready to push open this door and gain entry to "secrets" which had bested many other smart scientists and researchers for centuries before. And the world would never, ever be the same.

As with all paradigm shifts, it would become impossible to get the "natural selection genie" back into the bottle. Humans can never fully revert to the past unless modernity is lost through some calamity or else totally repressed and censored. The development of new ideas is commonplace and should be part of public knowledge. In science, the publication and open sharing of ideas is crucial to scientific advancement.

1859—The Pivotal Year that Changed Darwin and Humanity

On the Origin of Species by Means of Natural Selection, or the Preservation of Favoured Races in the Struggle for Life, would see its first edition printed on November 24, 1859, by the publishing company John Murray. The first pressing of the printed manuscript, all 1,250 copies, were purchased immediately. Darwin's book was an immediate best seller.[100]

The book was taken seriously, both as a form of intellectual liberation by those who supported Darwin and his theory and as a form of danger by his detractors. The worst religious detractors saw the book as sponsored by evil forces. Adam Sedgwick, one of Darwin's very early mentors and a pillar of modern geology, stood in deep opposition to natural selection for squarely religious reasons. He wrote:

> For you to deny causation…I call causation the will of God: and I can prove that He acts for the good of all His Creatures. He also acts by laws which we can study and comprehend.[101]

The publications *The New Englander* and *John Bull and Brittannia* each offered a scathing review of *Origin* on mainly theological grounds. And it was Minister John Leifchild who published in his anonymous review a refutation of natural selection that in part read:

> Theologians will say—and they have a right to be heard—why construct another elaborate theory to exclude Deity from renewed acts of creation? Why not admit that new species were introduced by the Creative energy of the Omnipotent.[102]

The Church was by this time squarely becoming tired of being told that neither human history nor the fate and destiny of the universe were controlled by a supernatural or all-loving deity. Darwin's theory would be the final shot into every theological precept. Since 1859, the

An 1871 unflattering caricature of Charles Darwin meant to ridicule the man and his theories

patient of organized religion has been on ontological life support and slips deeper into a coma with every scientific discovery made about the natural world.

But *Origin* received many positive reviews as well. It was clear right from the beginning that the debate over natural selection would draw lines in the cultural, scientific, and philosophical sands. Even today, the Answers in Genesis Ark Encounter and Creation Museum wouldn't keep their doors open if there weren't people who truly accept biblical creation as fact and natural selection as almost a form of sin to think or speak.

The publication *All the Year Round* offered a lukewarm if not positive review of Darwin's book. The *Chambers Journal* provided a favorable and almost poetic review. In December 1859 *The Saturday Review* offered a very detailed review, in part that read:

> In regard to that which is peculiar to Mr. Darwin's theory, we are far from thinking that the fruits of his labour and research will be useless to natural science. On the contrary, we are persuaded that natural selection must henceforward be admitted as the chief mode by which the structure of organized beings is modified in a state of nature.[103]

Clearly the power of advocacy was held strongly by each of the apostles. Darwin's critics were many, making him the whipping boy of the scientifically ignorant. In Darwin's defense, the apostles penned many articles and papers in support of him.

Each of Darwin's apostles had their own expertise and held varied research interests. Each debated the religious and scientific establishment, which in many cases was one in the same. They all had varying temperaments as well as different goals in widening Darwin's theory.

Huxley was bombastic in his debate style and writings, hoping to ultimately push theology out of science. Conversely, Asa Gray was more of a bridge builder, hoping to turn faith more easily toward the acceptance of natural selection.

What linked their literary defense of Darwin were three elements that ultimately qualified them for apostle-hood. The first was their friendship with Darwin and in many cases the friendships they had with each other. The second would certainly be a deep understanding that natural selection was not just a hypothesis but *the* answer to all forms of human, animal, and plant variation around the globe. They knew that denying this would essentially stop or significantly curtail our ability to make future discoveries and grow our culture and our understanding of science.

Finally, as experts in their own fields of research, they were able to debate and especially honor the defense of truth, a truth more than each could deny. The apostles knew that their ultimate legacy in fighting Darwin's fights in the debate halls and on the pages of popular publications was essentially to assure Darwin's legacy. And in doing so, they bolstered their own legacies as well.

Conclusions

In some parts of the globe the debate rages on as to whether natural selection is the single best theory we have to describe and define evolution. It remains a controversial hot topic and litmus test in politics, in our education system, and in personal ways when one comes up against a science denier. Especially when the denier is a member of one's own family or community.

There is only one formal place where natural selection is not "debated" any longer as a theory, and that is in scientific circles where trained and educated people work, teach, or otherwise gather. Another space *should be* the confines of the law. Judges have repeatedly struck down pieces of legislation that allow creationism and intelligent design to be taught as science in public schools. But state legislators continue to try to legitimize religion as science. There is no scientific "controversy," as evangelicals would love us to believe. It is a political wedge issue made purely to push faith over science.

The problems concerning evolution run deeper than some of the more well-known legal fights, such as those with school boards that wish to teach intelligent design or with local governments that want to place a cross on public grounds. About two-thirds of Americans accept the literal interpretation of the Bible or accept the idea that the universe and we humans in it evolved through the actions of a personal god. These religious ideas hark back almost three hundred years to Paley's divine watchmaker.

Ironically, in many cases the choices we are faced with in regard to accepting natural selection are the same ones people had back in 1859. We can either embrace science or abandon it for the supernatural. We can accept that we evolved or we can deny our long connection and common heritage to extinct ancestors. By association we can work to understand the operation of the natural world or we can ignore it at our own peril.

Darwin and Wallace lit the candle in the dark. It is up to each of us now to understand the importance of their observations and discoveries. In doing so, we can then light the way for others to follow.

CHAPTER 6

BIOGRAPHIES OF THE APOSTLES, AN INTRODUCTION

"I am trying to make clear through my writing something which I believe: that biography—history in general—can be literature in the deepest and highest sense of that term."

Robert Caro (1935–)
Journalist and Biographer

"The aim of science is to discover and illuminate truth. And that, I take it, is the aim of literature, whether biography or history…. It seems to me, then, that there can be no separate literature of science."

Rachel Carson (1907–1964)
Marine Biologist and Conservationist

It is true that all men and women are created equal. And in the most open and liberal societies, their talents and abilities can be fostered and developed to their fullest. In fact, one of the many things that equality offers to each of us is a starting point from which, if the circumstances are right, we can change the world. Usually this involves working in combination with other people.

Because humans are social primates with a long history of gathering and building, it should come as no surprise that during the European Enlightenment, scientists would begin to work together in many areas of knowledge to enhance our human understanding. From these collaborations our science would help us build new constructs of how we see ourselves in relation to our world and the cosmos.

The study of history allows us to hold a mirror up to our common humanity. We can marvel at our past behavior. Sometimes we learn from our mistakes and sometimes we repeat them over and over again. Sometimes we manage to skillfully avoid bettering our circumstances and sometimes we actually manage to better them. However, if we look closely and deeply enough at the patterns which make us human and highlight our success and failures, we can see common threads which should give us hope for the future, at least as they relate to collaboration within the social, behavioral, and physical sciences.

In modern times it is not unusual for a discovery to have many "parents." Indeed, with the widespread growth of information technology and collaboration between international scientists and their academic and government proxies, the use of new technologies fosters our understanding and demands for international focus collaboration.

Organizations such as the Rice Institute and Consultative Group for International Agricultural Research (CGIAR) have an international cadre of scientists who research and have invented sustainable agriculture to benefit the world's hungry. The Human Genome Project employed hundreds of geneticists from around the globe while it unlocked the chemical sequencing of the human genome. The World Meteorological Association, European Organization for Nuclear Research (CERN), the International Science and Technology Center, and the National Aeronautics and Space Administration (NASA) each have been working to understand the functioning of the planet and physical realities of the universe. And if we include the International Space Station (ISS) as another example of how science can build connections between people, it is clear that sometimes common scientific goals can build bridges where differing languages, cultures, faiths, diplomacy, economics, and social values have failed to bond us in the past.

Because collaboration was essential to Darwin's work, the purpose of the next five chapters is to offer the reader insights into the lives and scientific work of those who most closely surrounded and supported him prior to and after the 1859 publication of *Origin*. These men were essentially called by their friendship and admiration of Darwin to help make history by supporting one of the nineteenth century's most important naturalists and authors. It is clear that by knowing Darwin's most collegial relationships, we can gain a deeper understanding of how such friendships allowed this great thinker to develop his thoughts in collaboration with others, and in so doing to usher in modernity like few others ever will.

While the focus of this book centers on these five men of science who were ardent supporters of Darwin, it is clear that an intergenerational group of women were also in Darwin's corner. Certainly Darwin's wife, Emma, was instrumental in reading and editing most of Darwin's papers and books prior to their publication. His daughter

Henrietta would assist in editing *The Descent of Man*. And it would be his granddaughter, Nora Barlow, the child of Horace Darwin, who would fully restore her grandfather's full autobiography in 1958.

The apostles were like Darwin in that they were equally committed and deeply accomplished men. Each had his unique personal and family history and each was drawn to Darwin at different periods of his own scientific career. They all shared information at a time when new scientific knowledge and discoveries were being vetted in the academic societies and universities of Europe. At this time, new ideas and discoveries were being critiqued and debated as never before around the globe.

These biographies offer the reader brief backgrounds on Messieurs Huxley, Hooker, Gray, Draper, and Wallace. Though there are longer biographies of each, what we have concentrated on here is offering a brief history of each man, targeting his most significant contributions to science in general, to evolution in particular, and especially to Darwin and the group project of getting evolution accepted. All these men are important to the story of natural selection. Yet some are beginning to be lost to history while others are still well known and prominent. We hope that you enjoy their colorful lives.

CHAPTER 7

THOMAS HENRY HUXLEY

"Is Man so different from any of these Apes that he must form an order by himself? Or does he differ less from them than they differ from one another, and hence must take his place in the same order with them?"

Thomas Henry Huxley,
Evidence as to Man's Place in Nature

Briefly:

Thomas Henry Huxley was known to his contemporaries as "Darwin's Bulldog." Huxley was Darwin's staunchest supporter and most public advocate for the work of science and evolution. He was sort of the Christopher Hitchens of his time, lambasting and forcefully destroying anyone who would debate or call into question Darwin's work. His intellect and debating skills made him *the* catalyst for others to speak out in favor or Darwin's work and natural selection. His memorable 1860 debate against evolution foe Bishop Wilberforce, at which Huxley and Hooker reigned supreme, was germinal in making natural selection scientifically acceptable in the United Kingdom, in Europe, and across the Atlantic.

Huxley understood that although facts are said to speak for themselves, sometimes they need someone to do their shouting for them. Huxley was not afraid to do that shouting. Darwin was fortunate that he not only had a group of apostles, he also had one apostle in particular who could be a showman if he had to be. Although Huxley was a careful and rigorous scientist, he also knew how to use newspapers—the social media of their day—public lectures, open debate, and dramatic public dissections as ways of getting his points across.

Huxley also understood that facts must not be made to conform to religious dogma. This put him in opposition to the scientific establishment of his day, which was largely composed of wealthy amateurs and clergymen who were also naturalists. The concept of a professional scientist did not really exist in the time and place, and Huxley wished to change this and make science into a profession in its own right. That this has largely happened has been greatly beneficial to the progress of science, and for this we owe Huxley some thanks.

Huxley's Education and Life

Thomas Henry Huxley was a relatively small man in stature as an adult but he was a youngster who was a street fighter, with both the brains and determination to put down any adversary. He fought to raise himself out of extreme poverty, as he was born seventh of eight children, and his family had limited means of support. As an adult, Huxley fought for expanding educational opportunities for the poor and underserved, and he fought back hard on behalf of Charles Darwin and natural selection.

When Huxley was growing up, his family fell into that all-too-common Victorian pitfall of genteel poverty. His father had been a teacher, but when the school that he taught at closed, the family lost their source of income. Huxley's father was unable to find another position, so the family became impoverished.[104] This meant that although education was valued, the boys went out to work as soon as they were able. If the family had been aristocratic, the boys most likely would never have worked at any job or profession for money, except perhaps to join the military. If the family had been in the English working class, education beyond basic reading and writing might have seemed unworthy, and the idea of pursuing a profession might have seemed beyond reach. But since the Huxleys were middle class, the boys were sent out to earn money, and pursuing education in any way possible was a laudable—and practical—ambition.

Because of Huxley's upbringing, he did not like bullies and he was suspicious of authority. Fighting was in his nature and when those who could not protect themselves were harmed by the status quo or by powerful people not willing to help, he didn't just get angry, he fought back with words and deeds. So when Huxley saw the religious and social authorities of the day ready to pounce on Darwin, both before and after the publication of *Origin,* it was Huxley who led the battle for accepting natural selection in the face of self-serving and authoritarian ignorance and influence.

Huxley was born on May 4, 1825, in Middlesex, which is just outside of London. Then as now, his hometown of Ealing was considered a suburb of the greater London area. Huxley's father was a scientist and mathematician who taught in the early grades until his school was closed. This loss of employment sent the Huxleys from a relatively comfortable middle-class living into abject poverty. There were simply too many mouths to feed and little if no income to keep the family afloat.

This left little Thomas with few options of his own. He could stay in school and watch his family continue to decline socially and economically or he could do something about it and help support those he loved. At the age of ten Huxley quit school and began to take jobs to help support his family. As noted earlier, jobs for little boys at this time were dirty and dangerous. There were no labor laws and certainly no child labor laws to protect unskilled laborers and workers.

While the young Huxley suffered the slings and arrows of a family with few economic options, his interest in the world and a love of reading helped liberate the boy from the probability of life as an unskilled laborer. Huxley read everything he could get his hands on, from the classics to foreign language books, especially German, which he taught himself to read, write, and speak as a boy.

Huxley was also very fond of science textbooks. He read geography and geology books, as well as books on the description of fossils, which were making their way into in the hands of schools and lay readers of the time.[105] The young Huxley educated himself in the study of natural history, and in particular he learned about most species of invertebrates and vertebrates. His keen and encyclopedic mind stored and mapped the eccentricities of these creatures so well that he was considered an expert before the age of sixteen.

Since science was not a paying profession when Huxley was young, he did what many young men of a scientific bent did—he went into medicine. This is what Darwin himself tried to do, and what all the apostles did, with the exception of Wallace, who learned surveying.

How did he study medicine? Although the family's plunge into financial distress meant that he only attended school as a child from the ages of eight to ten, he read books on many subjects, including science, mechanical engineering, and drawing. His knowledge of German allowed him to read research papers about physiology that had been published in German, and his ability to draw helped with his later studies of zoology.

He studied anatomy. This was for practical reasons at first. But in so doing he became an expert and furthered his knowledge whenever possible. When evolution was being debated, his voluminous knowledge of comparative anatomy and his ability to do careful dissections really paid off.

He also served many medical apprenticeships. From the time he was about twelve, Huxley was employed by a series of physicians. First, he worked for his brother-in-law John Charles Cooke, a London doctor, and then for Thomas Chandler, another doctor. This gave him the opportunity to see close at hand how medicine, for both physical and mental illnesses, was practiced at that time.[106] This included seeing treatments that bear little resemblance to the way that medicine is practiced today, such as bloodletting, water vaporing, and even un-researched and near "magical" treatments and elixirs.

He also saw the ramshackle and sometimes rancid conditions in which most poor Londoners lived. Although his self-education during this time included learning physiology, physics, mathematics, and German, it also included learning about the deep poverty in which members of the working class in east London lived. Following yet another economic recession, Huxley saw first-hand the unemployment, starvation, and unsanitary conditions that the working class endured. In 1841 he was apprenticed to John Goodwin Salt, later changing his last name to Scott, another physician brother-in-law, this time in the north of London.

Essentially, Huxley was an orderly to these medical men. After years of these apprenticeships, he was able to use the knowledge that he gained in this way, plus the knowledge that he gained through self-teaching, to enter the medical field himself. As a mid-teen, Huxley won a medal in a contest at the Worshipful Society of Apothecaries, an early professional proving ground in the medical arts. All these things, along with a brilliant mind and nascent connections, provided him with the chance to study medicine at London's Charing Cross Hospital. Huxley had accomplished all this self-study, apprenticeship, and hard work by the age of sixteen.

Thus it was that in 1842 Huxley entered Charing Cross Hospital in London and obtained a small scholarship. At this point his career took off, in fits and starts.[107] In 1845 he published his first scientific paper, which described the membrane around the root of the human hair. This layer became known as Huxley's layer. That same year, at age twenty, he took and passed the first part of the exams required to get a Bachelor of Medicine degree. He did so well that he got the gold medal for anatomy and physiology.

It was clear that Huxley was on his way to an esteemed career in medicine. However, for reasons not clear, he did not show up for the second exam that was required for his degree. The result was that Huxley wound up without a degree, with his scholarship expired, and in debt, since he had borrowed money for living expenses. He would have been allowed to practice medicine without that degree but he was too young to apply for a license from the Royal College of Surgeons.

But in the strange way that Victorian England had, Huxley was able to pay his debts by signing on to a sailing expedition through an appointment to the Royal Navy. This enabled him to make money, further his zoological studies, and see some of the world. Since Huxley was a freethinker and scoffed all his life at social convention, his choices and interests would enlist him in some of the greatest debates in human history. What no one knew at the time was that Huxley would play a pivotal role in the how we view humanity's place in nature and how best to explain the natural world without the precepts of divine intervention.

But as a young man this longing for adventure simply overtook him and equally served as a way to pay back his debts. Huxley relinquished the opportunity for an easier and perhaps mundane route that could have been his if he had started a medical practice near London, where his personal and family connections were strong. Instead, he favored seeing the world and having more exciting experiences at sea. Because Huxley did not complete his studies, he became the assistant surgeon on the HMS Rattlesnake, whose purpose was to explore the South Pacific, including the British territories of Australia and New Guinea.

Thus, from 1846 to 1850, Huxley served on the Rattlesnake as it sailed to Australia and New Guinea for both discovery and surveying.[108] While on board, in addition to practicing medicine, Huxley also did research on natural history, and especially became interested in marine invertebrates such as medusae (jellyfish).

While at sea, Huxley sent most of his discoveries and the specimens they represented back to England. In 1849, he published a paper on these organisms. It showed how adult jellyfish had tissue layers that were similar to the tissue layers that exist at one point during vertebrate embryological development.

These shipments and Huxley's commentaries on his findings brought him into the circle of naturalists, which included both Lyell and Hooker. It was just a short leap from there that he would meet and begin corresponding with Darwin by the early 1850's. This would lead to a lifetime of letter writing and friendship between the two men.

When the ship returned to England, Huxley was granted leave so that he could prepare scientific reports from his voyage. He slowly became famous as a scientist, was elected to the Royal Society of London, and received the Royal Society's Gold Medal.

However, his financial life was not as golden as his scientific life. To make matters more complicated, Huxley refused to return to his ship when his leave was over, and wound up being discharged from the Royal Navy.[109] Fortunately, in 1854, Huxley got a job as a professor of natural history at the Royal School of Mines in London. The next year he additionally became the Curator of Fossils at the Museum of Practical Geology in London. There he began a series of lectures for workingmen.

As Huxley had known real poverty as a child, and it was certainly stressful for him to be a man well admired but without the funds to sustain himself, he was certainly stressed post-*Rattlesnake*. In this way, he and Asa Gray were kindred spirits, both brilliant outsiders who, because of their lack of influential family connections, had to truly earn their place in the scientific communities where they had staked their professional and personal lives.

Huxley was a brilliant paleontologist. But his early paleontological work, which consisted of assumptions on extinction and the number of species which had lived in the past, was actually incorrect. One of his saving graces, however, was his ability to draw the fine details of fossils and then connect them to earlier species. Indeed, his later work on the evolution of coelacanths and of the common horse, as well as his description of the flying dinosaur archaeopteryx, still impacts what we know about species leaving the ocean, vertebrate evolution, and the bird-dinosaur connection.

Huxley's career was now in full swing. He gave popular science lectures, developed his own course at the Royal School of Mines, helped to organize the collections at the Museum of Practical Geology, and wrote a regular science column for the *Westminster Review*.

During this time, his research interests moved from invertebrates to vertebrates, fossils, and paleontology. In 1856 he received a Fullerian lectureship at the Royal Institution of London.[110]

During this period, good luck turned to Huxley and it turned to him full blast. While he was teaching at the School of Mines, Huxley also obtained a simultaneous academic chair at the Royal College of Surgeons. Huxley also worked as a popularizer of science, scheduling and planning numerous public lectures for those with an interest in science. In this way he was trying to educate those who would not otherwise have the opportunity to know or understand all the changes and findings of the science of the day.

This career growth and stability also brought his beloved from Australia. While serving on the *Rattlesnake,* Huxley had fallen madly in love with one Henrietta (Nettie) Heathorn. Nettie was the daughter of a pub owner. But even the frugal Huxley could not hope to promise her a good life in England unless he had a stable job and sufficient income. After an eight-year and very long-distance engagement, Thomas and Nettie were married in London in the year 1855.

Huxley's other scientific research also flourished during this time. Crucially, during the late 1850s, but before *Origin* had been published, Huxley began doing research and publishing papers on vertebrate embryology. He found that in classifying organisms, it was important to consider not just their adult structures but also the development of these bodily structures during embryological development.

The idea of classifying organisms according to common structures predates the idea of evolution by natural selection, having been put into common scientific use by Linnaeus. As an example of structures that animals have in common, humans have two easily-identified bones in their forearms, and one bone in their upper arm. Birds, whales, bats, dogs, and horses have the same bones, still easily-identified but modified for their different forms of locomotion. This is known as the principle of homology. Cuvier expanded this system of classification by common structure to include fossil organisms. Huxley then expanded it further to include structures that develop similarly prior to birth. In all these cases, the process of classifying organisms in this way edges the classifier ever closer to asking the question of "Why are these structures so similar?" and "Could it be that these organisms are related?"

We now know that when organisms have structures like these in common, this likely means that they share a common ancestor.

Richard Owen Gets Involved

Richard Owen was a contemporary of Huxley and Darwin who was a zoologist of great importance, since he was the head of the natural history department at the British Museum. The British Museum, with access to specimens obtained throughout Britain's far-flung empire at the height of its power, had one of the best natural history collections in the entire world.

Richard Owen disagreed with elements of Linnaeus's classification scheme and actually tried to reverse Linnaeus's well-accepted idea that humans should be classified with other primates. In fact, he claimed that humans were so very different from all other animals that they should barely even be considered to be mammals, much less primates.

He based these grand claims on the idea that human brains were entirely different from ape brains. He saw them as different based on three very specific and to him undeniably important features:

1) That only human brains had a posterior lobe that stuck out at the back of the brain even beyond the structure called the cerebellum.

2) That an opening in the interior of the brain known as the lateral ventricle had an additional extension on it that was unique to humans. This extension was called the posterior horn (or posterior cornu) of the lateral ventricle.

3) That only human brains had an *additional extension* off of the posterior horn of the lateral ventricle that was called the hippocampus minor (known today as the calcar avis).

It may seem odd that minor variations in the cavities within the brain caused such excitement, but such was the case. Owen also made a more general claim that the outermost layer of the brain, the cerebral cortex, had deeper and more numerous convolutions in humans than in apes.

If all this seems like a slender thread on which to hang an entire reclassification scheme, it is. But many people are simply resistant to the idea that humans are closely related to apes and will cling to anything that appears to offer them a way out. Owen was one of those people, and there are many like him today.

Owen admitted that apes and humans have a vast amount of their anatomy in common. But, he insisted, those three tiny differences between our brains and those of apes were so vast in terms of what they enabled humans to do that they were sufficient justification for this radical reclassification.[111]

If only those differences had actually existed!

Unfortunately for Mr. Owen, not only did those differences not exist, he also crossed swords with Huxley when he claimed that they did.

The story begins in the 1850s, when Richard Owen started making these claims that human brains are unique in ways so fundamental that they could not possibly be classified with apes and other primates.

Then he went to the even more extreme position. In 1857 Owen presented a paper to the Linnean Society stating that humans were so different from all other animals that they were actually a different suborder of mammals—that is, not only were they not primates at all, they were barely even to be classified as mammals. The differences he cited were the three given above. He based this claim on three drawings—one that he had done himself of the brain of a South American monkey, one done of a human brain by the German anatomist Friedrich Tiedmann, and one done of a chimpanzee's brain by the Dutch anatomists Jacobus Schroeder van der Kolk and Willem Vrolik. This paper, "On the Characters, Principles of Division, and Primary Groups of the Class Mammalia," was published in the *Proceedings of the Linnean Society of London* in 1858.[112]

Owen said of the human brain:

> [The] posterior development is so marked, that anatomists have assigned to that part the character of a third lobe; it is peculiar to the genus *Homo*, and equally peculiar is the 'posterior horn of the lateral ventricle', and the 'hippocampus minor', which characterize the hind lobe of each hemisphere. The superficial grey matter of the cerebrum, through the number and depth of the convolutions, attains its maximum of extent in Man.[113]

So there are Owen's three specific claims, and the one more general one that I explained earlier.

With those claims, Owen delivered himself into Huxley's merciless hands. Huxley started to quietly dissect monkey brains. He was determined that

> before I have done with that mendacious humbug I will nail him out, like a kite to a barn door, an example to all evil doers.[114]

Huxley did not present his findings about monkey brains immediately, and instead concentrated on weakening Owen's reputation in other ways. So, for instance, Huxley gave a public lecture in June of 1858 in which he directly challenged another of Owen's other pet theories—with Owen in the audience!

Meanwhile, Darwin had been corresponding with Huxley about natural selection since 1854. Darwin had seen Huxley's extremely negative review of *Vestiges of the Natural History of Creation* in 1854 and decided that it would be wise to quietly approach this firebrand and present his own theory of natural selection privately to him. Huxley and Darwin wrote back and forth, and Darwin slowly presented his evidence to Huxley.

Although Huxley at first had reservations about specific aspects of Darwin's view of natural selection, particularly Darwin's emphasis on gradualism, he nonetheless knew that the basics of evolution were correct. Darwin's mountains of evidence could not be denied.

Huxley even prepared the ground for Darwin by writing two pieces that suggested simultaneously that biblical notions of how species originated were probably wrong, while also saying that the Lamarckian style of evolution, which included that species could consciously change themselves by reaching higher or swimming faster, as was hinted at in *Vestiges of the Natural History of Creation*; was also wrong.

Then on November 24, 1859, *On the Origin of Species* was published by Charles Darwin and both men's lives were changed forever.

Family Life

Meanwhile, in realms other than science, Huxley had a large and loving family. Though it was not uncommon for people with means to have large families at that time, Huxley and his wife were prodigious. The Huxley family eventually included eight children and many grandchildren. Some of his descendants became well known architects, intellectuals, authors, and humanist celebrities in their own right.

Sadly, the Huxleys first son Noel only lived till the age of four and passed away in 1860. Two daughters, however, Jessie and Marian, followed, lived to adulthood, and eventually married. As with Darwin, Draper, and Hooker, each man would lose a favored child early in life to the dreaded diseases of the day. Noel's loss most likely made Huxley think he needed to have a large family simply so that he would have some children who survived.

Huxley's Career after *Origin* Was Published

At this point, the more colorful and pugnacious aspects of Huxley's personality came into play. He wrote to Darwin the day before *Origin* was officially published and said:

> I trust you will not allow yourself to be in any way disgusted or annoyed by the considerable abuse & misrepresentation which unless I greatly mistake is in store for you— Depend upon it you have earned the lasting gratitude of all thoughtful men—And as to the curs which will bark & yelp—you must recollect that some of your friends at any rate are endowed with an amount of combativeness which (though you have often & justly rebuked it) may stand you in good stead.

I am sharpening up my claws & beak in readiness.[115]

Then he had a stroke of luck. A Mr. Lucas had been asked to write a review of *Origin* for *The Times* of London, but he didn't have the appropriate knowledge of science to do so. After asking around, he decided to ask Huxley to do the review! Huxley's positive, skillful, and anonymous five thousand word review filled three and a half columns and was printed in full, with Mr. Lucas providing an introductory paragraph or two. Since the *Times* was the leading shaper of public opinion in that day, the fight for evolution was off to a good start.

Newspapers at the time were not just news sources. They also played the role that social media do today. To begin, many were openly partisan. They sometimes published articles and reviews anonymously, and writers sometimes used pseudonyms. Content was sometimes provided for free in order to promote certain points of view or to give exposure to various people who wished to become famous. Does this sound familiar? Newspapers could also be gossipy and personal. Many came out twice a day, and a big city could have a dozen or more different newspapers and magazines.

What's more, rather like social media today, people commented back and forth on issues of the day through letters to the editor, with letters sometimes being as long as articles. (There's that free content again!) The letters at times were both heated and personal. Slanging matches, as these were known, drove circulation in much the way that provocative posts on social media can go viral, thereby benefitting their platforms. So authors would write, commenters would comment, the public would buy the newspapers, and the publishers would make money.

Twitter wars are today's version of the Victorian era's printed slanging matches.

Huxley's favorable review of *Origin* was then re-published in *Gardner's Chronicle*, which led to the exchange with Patrick Matthew that was discussed in the Antecedents chapter. Many papers and magazines came out as opposed to Darwin, but the overall result was that evolution was being discussed far and wide. This allowed Huxley to then write articles in *Macmillan's Magazine* and the *Westminster Review* about the general uproar.

In commenting, Huxley said that "old ladies, of both sexes, consider it a decidedly dangerous book."[116]

Richard Owen, from his high perch at the British Museum, was vocally opposed to evolution and he made the fatal mistake of engaging in slanging matches with Huxley in the newspapers on that subject. Huxley not only had all the evidence on his side, he was positively gleeful in taking Owen down. Owen's related fatal mistake was engaging Huxley on the battlefield of vertebrate anatomy, a subject that Huxley knew far better than Owen did.

At this point all the research that Huxley and others had been doing regarding the true anatomy of primate brains came into play.

> Think back to that famous 1860 meeting of the British Association for the Advancement of Science in Oxford. On Saturday, Huxley entered that debate already annoyed. What had happened earlier to get him so irritated? The record shows that on the preceding Thursday there had been a paper presented that was titled "On the Final Causes of the Sexuality of Plants with Particular Reference to Mr. Darwin's Work" by Charles Daubeny. After that presentation the chairman asked Huxley if he had any comments. But Huxley still thought, at that moment, that a public venue was the wrong forum for such a serious discussion. Then Owen spoke up and claimed that he knew of facts that would convince the public that Darwin's theory was untrue. He then repeated his fateful claim that the brain of the gorilla was more different from that of man than from that of the lowest primate particularly because only man had a posterior lobe, a posterior horn, and a hippocampus minor.

This qualifies as yet another instance in the history of science when someone's lovely idea was destroyed by an ugly fact. At that presentation, Huxley firmly but politely denied Owen's claims, cited previous studies, and promised to provide support for what he said.

However, this meant that Huxley was already tired of nonsense before he even entered the hall where the debate took place on Saturday. And being annoyed at Owen and being annoyed at Wilberforce amounted to much the same thing—they were both bent on denying evolution, weren't ashamed of being articulate while spouting nonsense, and were friends with one another. So when faced with a bishop who spouted articulate nonsense, and also faced with a noisy audience that was uninterested in a calm discussion of facts, Huxley let the bishop have it with both barrels, and sat down. But he also quietly continued to accumulate anatomical evidence to use against Owen.

The Death of a Child Spurred Still Greater Passion

Huxley would undoubtedly have carried on with debunking Owen. But his son Noel died of scarlet fever in September of 1860. He was devastated by the loss. However, with encouragement from his friend Charles Kingsley, he managed to channel his anguish into fury at Owen.

1861—Huxley and Owen Mix It Up

In 1861 Huxley and Owen ramped up their attacks and counter-attacks.

In January 1861 Huxley re-launched the *Natural History Review* and published "On the Zoological Relations of Man with the Lower Animals." In this article, he attacked Owen's three claims about three features of the brain that he considered to be unique to humans, using letters, citations, and quotations from other anatomists.

Huxley's refutations were thorough, cited Owen himself, and strongly suggested that Owen was not merely wrong but knew that he was wrong and repeated his lies anyway.

In addition, the first two issues of *Natural History Review* had five major articles by other scientists that also attacked Owen's anatomical claims.

> From February to May, Huxley delivered a series of lectures for working men at the School of Mines where he taught. The series was titled "The Relation of Man to the Rest of the Animal Kingdom." The lectures were very popular and Huxley told his wife, "My working men stick by me wonderfully, the house being fuller than ever last night. By next Friday evening they will all be convinced that they are monkeys."[117]

Meanwhile, gorillas became a hot topic in London when the explorer Paul Du Chaillu returned to London. Owen arranged for Du Chaillu to speak at the Royal Geographical Society on February 25. The event was spectacular and allowed Du Chaillu to not only speak but also display his collections. Recall again that collections were of great importance to Victorians.

Then on March 19 Owen followed this up by giving a talk himself about brains, titled "The Gorilla and the Negro." In this lecture, he deliberately compared the supposed brain anatomy of those who he considered to be the "lowest" humans ("the Negro") with that of a gorilla. In comparing them, he muddied the water about his three claims of three indisputable and tremendously important differences between human brains and all other primate brains. He said that the dispute was not a matter of facts but a matter of interpretation. So Owen's slender thread of anatomical differences became even thinner, since he was no longer willing to dispute the facts. Except for one fact, which he hedged carefully. He said that only humans had a hippocampus minor "as defined by human anatomy," which is a circular sort of definition capable of muddying water even further. So his entire claim that humans are not a type of primate, and are barely even a type of mammal, went from being based on claims of tiny anatomical differences to being based on "interpretations" of tiny differences that did not exist anatomically.

On March 23, this lecture was published in the *Athenæum*. Unfortunately, it had illustrations accompanying it that were both inaccurate and unlabeled. From his lofty position at the British Museum, perhaps Owen thought that this did not matter. Or perhaps he couldn't do any better.

In the next week's issue of the *Athenæum*, Huxley responded with an article titled "Man and Apes" in which he made fun of Owen's lousy illustrations and his failure to notice the findings of other anatomists, who said that the three structures were indeed present in primates besides humans. The *Athenæum* republished "Man and Apes" on Sept. 21, 1861, which served to further Huxley's work and advocacy.

The week after that, Owen published a letter in the *Athenæum* blaming "the Artist" for the poor illustrations, but he still claimed that his basic arguments were correct, and referred back to his 1858 paper in the *Proceedings of the Linnean Society of London* titled "On the Characters, Principles of Division, and Primary Groups of the Class Mammalia."

This was the paper in which he made his three grand claims about the human brain—that is, that only human brains had a hippocampus minor, that only human brains had a posterior lobe that stuck out at the back of the brain, and that only human brains had a posterior horn of the lateral ventricle.

The exact quote: "[Archencephala's] posterior development is so marked, that anatomists have assigned to that part the character of a third lobe; it is peculiar to the genus *Homo*, and equally peculiar is the 'posterior horn of the lateral ventricle', and the 'hippocampus minor', which characterize the hind lobe of each hemisphere. The superficial grey matter of the cerebrum, through the number and depth of the convolutions, attains it maximum of extent in Man."

Then on April 13, Huxley's published response was to say, "Life is too short to occupy oneself with the slaying of the slain more than once."[118]

The authors of this book agree that slaying the already slain can be very tedious, but the history of evolutionary biology is one of repeatedly slaying the already slain, only to have others claim of the corpse that "he's not dead yet!"

Darwin, meanwhile, had the fun of reading these slanging matches from the comfort and safety of Down House.

The year 1861 also saw Huxley's specific, well-explained, well-illustrated, carefully scientific, and crushing response to Owen, titled "On the Brain of Ateles Paniscus," published in the *Proceedings of the Zoological Society of London*. In this article, Huxley first pointed out the work done by other scientists who had found the infamous three structures in the brains of both chimpanzees and orangutans. He then deliberately switched to discussing monkeys—which are much less like human beings than apes, like chimpanzees and orangutans, are. *Ateles paniscus* is a species of new world spider monkey.

This was an intentional put down. Huxley was making it clear that Owen was so mistaken about brain anatomy that not only do humanity's *near* relatives possess all of Owen's seemingly magic brain structures, their more distant ones do too! And sure enough, in the course of Huxley's article, he describes in careful, well-illustrated, well-labeled detail the structure of *Ateles paniscus*'s brain, based on dissections done by Huxley himself. The large posterior lobe that sticks out the back, the posterior horn of the lateral ventricle, and the hippocampus minor were all found in spider monkey brains by Huxley himself. (In case anyone is worried that Huxley might have cheated, he had another trick up his sleeve that came out in 1862.)

Huxley further pointed out that other scientists had found that a well-developed posterior lobe, far from being exclusive to human brains, is actually largest and sticks out behind the farthest in squirrel monkeys—a fact that had been published in 1855! To rub it in a bit more, Huxley pointed out that capuchin monkeys and chimpanzees also have posterior lobes that are actually larger than those found in humans.

Huxley further related that both the posterior horn of the lateral ventricle *and* the hippocampus minor were relatively larger in spider monkeys than in humans. One can practically hear Huxley saying, "Take *that*, Owen!" Instead of course, he wraps up the paper by discussing all the other structures in the spider monkey brain, as a good scientist should.

During 1861, Huxley also published five scientific works of his own that were not related to Darwin's work. These were on subjects ranging from fossil fish to fossil ungulates (animals like horses and tapirs and rhinoceroses).

1862—William Henry Flower Takes a Star Turn

Other scientists continued to be a part of the fray. In 1862 the anatomist William Henry Flower—a protégé of Huxley's—published a paper in the venerable *Philosophical Transactions of the Royal Society* called "On the posterior lobes of the cerebrum of the quadrumana." In this article, he refuted Owen's three claims no less than sixteen times each—looking at sixteen different primate species, from prosimians through apes, as well as reviewing other scientists' work.

However, his best dramatic moment came at the 1862 meeting of the British Association. It is not often in annals of science that scientists ambush other scientists at a scientific meeting, but this was one of those times, and it was probably the best one. Remember that Huxley had publicly disagreed with Owen's claims at the 1860 meeting regarding Owen's three claims of the incredible brain differences, and had promised to provide published scientific studies to back up what he said. For the past two years Huxley had done so, yet Owen had never stopped pressing his claims. At the 1862 meeting, Owen was still at it, repeating those tired assertions.

But this time he got ambushed. At the end of Owen's session Flower stood up and made the following remarkable announcement:

I happen to have in my pocket a monkey's brain.[119]

He did, too! He then proceeded to do a public dissection of said brain, allowing all the assembled scientists, and therefore all the rest of the world, to see in real time that that the brain indeed had all the parts that Owen said that it ought to be missing. Huxley and Flower were close associates, and it is entirely likely that Huxley put Flower up to this.[120]

Things Settle Down

In 1863 Huxley published *Evidence as to Man's Place in Nature*. In this extensive essay he compared at length the anatomical similarities between humans and apes, including brain evidence, skeletal evidence, and fossil evidence. He also included a six-page section describing how Owen had "suppressed" the truth about the hippocampus minor.

He basically accused Owen of lying, saying that the issue was now one of "personal veracity." Charles Lyell came down strongly on Hux-

ley's side, publishing "The Antiquity of Man."[121] This pretty much settled the matter among scientists, and is especially laudable since Lyell
was not yet at that point fully convinced about evolution by natural
selection. Nonetheless, he was willing to stand up for scientific truth
when he saw it.

By 1866, Owen was trying to have it both ways, simultaneously admitting that all structures in the human brain were also present in apes'
brains, but somehow still insisting that they only existed "under modified form and low grades of development," and further insisting that
none of this in any way changed his assertion that humans should be
classified as a separate subclass of mammals, separate from apes and all
other primates. He threw in some nasty comments about Huxley and
other brain anatomists as well, calling them "puerile," "ridiculous," and
"disgraceful."[122]

He never got his reclassification, never succeeded in ruining Huxley's reputation, and the great "hippocampus question" became a weird
epic circus in the history of science. The hippocampus minor, now
known again as the calcar avis, continues to exist, but only as a minor
part of the brain, which is where it belongs.

But Not Before the Public Has a Laugh

The Great Hippocampus Question found its way into popular comedy. How could it not when two famous scientists were having the
equivalent of Twitter wars about it in the popular press? The satirical magazine *Punch* published a poem, "Am I a Man or a Brother," in
1861 about the controversy and a cartoon about Huxley and Owen
in 1865.

The possibility of Huxley grappling with Owen remained a public curiosity even as late as 1865. Seen here in the publication *Punch*,
which offered their take on the match. Their vision of the British Association for the Advancement of Science looked like a tag-team event
which not only included Huxley and Owen but John Tyndall, Charles
Lyell, and even Michael Faraday.

The 1862 British Association meeting provided inspiration for
Charles Kingsley's satirical skit titled "*Speech of Lord Dundreary in
Section D, on Friday Last, On the Great Hippocampus Question.*" Lord
Dundreary was a stock character in British comedy at the time—an
aristocratic twit with a habit of mangling proverbs.

An anonymous pamphlet published in 1865 provides a fictional account of Huxley and Owen being tried in court for brawling in the

SEPTEMBER 23, 1865.]   PUNCH, OR THE LONDON CHARIVARI.   113

THE BRITISH ASSOCIATION.

streets. Most famously, in 1863 they became characters in a children's book called *Water Babies*, and the great controversy became known as "the great hippopotamus test." In it, a child declares that he is puzzled by the strange things being said at British Association meetings. He says that he had thought that the differences between him and apes were things like:

> "being able to speak, and make machines, and know right from wrong, and say your prayers" Instead it appears that the difference is having "a hippopotamus major in your brain." He now knows that "if a hippopotamus major is ever discovered in one single ape's brain, nothing will save your great-great-great-great-great-great-great-great-great-great-greater-greatest-grandmother from having been an ape too."[123]

Separating Church and Science—The X-Club and *Lux Mundi*

One of Huxley's motivations in defending evolution so vigorously was his desire to get religion out of science. Until that time, English society in general was happy to have scientific research done by clergy who had time on their hands. Clergy were seen as educated men, and therefore capable of the thought work necessary for science. If a clergyman's job of ministering to his congregation didn't take up all his time, then wandering around in fields collecting specimens and trying to figure out how God's work fitted into their findings seemed like a harmless enough pastime. The same was true for wealthy amateurs with no profession and time on their hands.

Remember that Darwin himself was trained as a minister, and would have become one if he hadn't shipped out on the *Beagle*. There were a few positions for trained professional astronomers at the Royal Observatory in Greenwich, but there were very few other professional scientific positions. This is why nearly all of Darwin's apostles trained as physicians, even the botanists. Medicine was one of the few professional fields that had a hint of science in it.

Getting religion out of science is necessary in order for science to succeed. If a scientific question is met with the answer that "God did it," then no further research is necessary, and real answers will never be found. Science only progresses when supernatural answers are rejected, and more work is done to find material causes for material phenomena. And marvelously, when scientists do reject supernatural answers and look for natural ones, eventually they are found, and tested. After those answers are tested enough, they become useful scientific knowledge and the boundaries of what is known get pushed back a little bit further.

Another problem is that religion requires that certain conclusions must be reached. If it is a foregone conclusion that the biblical story of creation is true, then all information in geology, geography, biology, astronomy, and physics must be made to somehow fit into this already-known answer, even if the resulting explanations become increasingly untenable and farfetched.

Huxley understood this, and understood that religion, including the then-powerful Church of England, had to be made to leave science alone. He also understood that scientists should not occupy the social role of monks, living loveless and childless lives devoted only to knowledge. He wanted science to be a profession, and scientists to have the status and salaries of professionals.

And Huxley, vocal and brilliant though he was, also understood that he couldn't do it alone. For this reason he became a founding member of the X Club.[124][125] This club was intended to bring together elite like-minded people from all intellectual, governmental, and financial walks of life to help inform British society, advise the government and monarchy, and to uplift and change English culture in ways to equalize and intellectualize the masses. Think of the X Club as Huxley's version of the Illuminati: a group of men trying to manipulate the destiny of the world through their powerful connections, ideas, and other resources.

Each Thursday the members and their guests would have dinner and discuss issues of the day and attempt to draw consensus on how they, as a group, would respond if asked (and at times when not asked). One of the living legacies of the X Club is the science publication *Nature*, which began as a literary voice for the group in 1869 and continues to serve as one of the foremost science journals still published.

The members of this club firmly believed that the natural world was governed by causes and effects that can be known by science. They also believed that science had many practical benefits for a society, is extremely beneficial as a mental and educational discipline, and leads to reliable knowledge of the natural world.

They claimed on this basis that cultural leadership should go to scientists rather than clergy. They specifically defended evolution by natural selection and Charles Darwin himself. And they campaigned for science education at all levels, jobs for scientists, and government support for science. They largely succeeded. Members of the group, including Huxley, gained intellectual and social prominence and power, advised the government, did vast amounts of educational outreach, and in Huxley's case wrote biology textbooks and revised the British educational system at all levels so that it would include science education.

However, among all these practical signs of success, perhaps the most telling cultural evidence of their reach and success is this: in 1889 the book *Lux Mundi* was published. This collection of essays was done by influential clergy in the Church of England and edited, ironically, by a future Bishop of Oxford. *Lux Mundi*, which means "Light of the World," assumes as its basis that evolution by natural selection is true. The essays seek, sometimes explicitly, to explain and even recast the faith in terms of Darwinian evolutionary theory.

So the mighty Church of England gave way to science. The religion that stretched all over the world courtesy of the British empire, the religion that had only stopped executing people for heresy in the seventeenth century, changed its supposedly timeless truths in order to make way for the real thing.

Darwin's Apostles had succeeded.

Huxley's Later Career

In 1870, Huxley's focus would change a bit as he was appointed to the London School Board. He worked to transform public education so that all children, especially those with little means, could become informed modern citizens.

Huxley would hold the title of chair at the Royal School of Mines for slightly more than thirty years, resigning in 1885, a decade before his death. He would also hold numerous other simultaneous professorships and presidencies at some of the most prestigious British academic institutions and professional societies.

Huxley was also a man who was deeply loyal to and served in government in numerous capacities for the sake of the public trust. In over twenty years he either led or took a major role in nearly ten Royal Commissions, developing governmental policies to support science education and to consult with Parliament on matters pertaining to science, public education, and issues concerning the interpretation of the natural world. In all of these positions, Huxley served as unstoppable advocate and essentially sent amicus briefs to the powerful about his ideas.

For the rest of his life, Huxley continued to promote evolution while simultaneously doing his own research, doing public lectures for working people, and serving on multiple boards. These boards took up vast amounts of his time and were devoted to reforming and improving multiple aspects of English society, from government policy through education.

Education Reform

For primary school, he promoted what is now our basic, broad instruction. That is, we are expected to receive, whether in public or private schools, a broad education in reading, writing, arithmetic, art, science, and music. This may seem like a common-sense approach to a young person's education now, but this type of education was nearly unobtainable in England when Huxley was young.

Huxley's views of education were very modern and would seem familiar today. This is largely because the curriculum he helped to promote is still in use. He worked to ensure that all children would learn to read, write, do math, and understand science; but he viewed a well-rounded education as one that also included learning the humanities, art, and music. Ironically, Huxley the agnostic found that having students read the Bible was helpful too, as he described it building ethics and moral character. His goal, though, was to rewrite the Bible removing the religious dogma, the Christian theology which he deeply opposed, and mysticism—ensuring that only the fables of good and evil remained.[126]

For post-secondary education, he advocated for the type of liberal arts curriculum now used by nearly every American college. That is, two years of basic liberal studies followed by two years of higher-level, more specialized work. In Oxford and Cambridge during Huxley's time, education was focused on the classics.

Not surprisingly, Huxley thought that education in the sciences should be a part of the available curriculum. But he thought that learning done only through textbooks was a waste of time. He emphasized the value of learning by doing and through direct observation. He expected his students to spend many hours in laboratories, dissecting and directly observing all aspects of vertebrate anatomy.

One remarkable aspect of this was Huxley's own distrust of received wisdom. Early in his career he had routinely discovered that "facts" as printed in textbooks were simply wrong. He found this out by doing his own careful examinations. We can see how these early experiences propelled him to both accept and promote evolution by natural selection. On the one hand, Darwin too was willing to overturn received wisdom. But on the other, Darwin also provided mountains of carefully gleaned scientific evidence to back up his then-radical theory. For Huxley this was a combination that was bound to be fascinating.

The downside of Huxley's emphasis on learning by doing was that, since he was an anatomist, he had biology students learn biology through anatomical dissection, largely to the exclusion of everything else. A student at Cambridge in 1880 remarked:

> We were not so very far off from the *Origin of Species*, and we were even contemporaneous with Darwin's later works, all of which dealt with living creatures, living organisms, and yet our obsession was with the dead, with bodies beautifully preserved and cut into the most refined slices, stained in various pigments.[127]

This meant that the more modern fields of biology such as ecology, ethology, and molecular biology perhaps came along later than they might have otherwise, and it made biology less interesting to generations of students. On the one hand, learning by doing dissection is indeed learning by direct observation rather than by reading and accepting facts without verifying them. But on the other, biology is a much wider and more interesting field than can be understood through dissection alone.

Travel to America

In 1876 Huxley went on a long tour of the United States. Landing in New York on August 5, he promoted evolution whenever he could, playing to packed lecture halls across the eastern half of the country. He traveled more than three thousand miles and got plenty of publicity. He socialized with Governor Porter of Tennessee and found a bust of himself at Vanderbilt University. He wowed audiences with news of fossil discoveries that had been found right in the United States. In New York he not only had capacity audiences for three days but he also wound up with a commemorative issue of the *Tribune* in his honor, containing the full texts of his New York speeches.[128]

He also had the fabulous opportunity to meet with American scientists and see American fossil collections. He declared that seeing the collection of Badland fossils in the Peabody collection at Yale was by itself "worth all the journey across." He saw the famous three-toed tracks on the banks of the Connecticut river that are now known to be dinosaur tracks, and saw the collection of fossils that showed the early evolution of the horse. This he declared was "the most wonderful thing I ever saw."[129]

So inspired was Huxley by the fossil remains of early horses that he sketched an imaginary tiny "dawn man" ("Eohomo") riding on a dog-sized then-hypothetical "Dawn horse" (*Eohippus*) that was based on the fossils present in the Yale collection.

Huxley gave this sketch to the man who had collected the fossils, Dr. O. C. Marsh of Yale University. In reality, early humans and early horses missed each other by fifty million years. Huxley knew this, but the sketch is fun anyway.

By the time he left the United States, Huxley had many more Americans understanding that you can't doubt evolution without doubting science itself.

Huxley's sketch of Eohippus being ridden by "Eohomo." The original drawing is still in the Yale Peabody Museum Archives.

More Evolution Promotion and Combat

One of the bolder projects that Huxley ever engaged in was a series of printed debates with William Gladstone about the value of evolutionary biology versus the value of knowledge of the Bible.

These were printed in a review titled *The Nineteenth Century*, and lasted from 1885 through 1891. Considering that Gladstone was prime minister of England four different times between 1868 and 1894, this meant that Huxley *started* this debate when the man had already been prime minister, was still active in politics, and as it turned out, would be prime minister again. Thus Huxley was debating the value of scripture versus evolutionary biology with the most powerful politician in his country at the time. This alone was daring enough. But not surprisingly, Huxley didn't pull any punches against so powerful a man. Instead, by most accounts, Huxley ran rings around the statesman. But regardless of who came out best in any given exchange, the basic fact is that Huxley managed to engage one of the most prominent politicians of his era on this topic and combated him as an equal, thereby putting science and evolution front and center as a force to be reckoned with in the public eye.

During this period, Huxley continued his lectures on matters relating to evolution in his "Lectures to Working Men" at the Royal School of Mines.

Reforming Great Britain

During his later years as an established scientist, Huxley spent many hours serving on committees and commissions of various sorts, trying to help or reform things from Indian agriculture to Scottish fisheries to contagious diseases to education.

What these have in common, besides showing how extraordinarily hard Huxley worked throughout his life, is that they show Huxley's willingness to make the world a better place, including by doing the unglamorous work of committees and commissions, as well as doing the more flamboyant writing and appearances for which he is better known.

What we can see in this is that Huxley was a true radical, in the etymological sense of the word. Radical means "uprooting," as in uprooting what has been done, thought, and accepted and planting in something else.

With his willingness to engage in public intellectual combat, and even his delight in it, Huxley might be suspected of being a mere gadfly. Certainly for many years he deliberately engaged in combat with many well-connected people, and did his best to put an end to ideas and habits of mind with which he disagreed. But a mere gadfly would be content to buzz. Huxley, as he became a well-connected insider himself, made a point of using his insider status to change society for the better.

When Huxley did retire in 1885, he had little choice. The man had spent six decades of his life fighting for himself, for others, for science, for public education, and to give the world standards in medical and scientific training that were wholly different from his own apprenticeship.

Huxley did suffer emotional breakdowns throughout his life, and this funk and anxiety essentially took him out of most of his duties several times in his career.[130] Clinical depression had dogged him on numerous occasions, including as a child medical apprentice, aboard the *Rattlesnake,* and prior to gaining the position as educator in the Royal School of Mines. It was an illness which seemed to have some biological basis in the family as his father, brothers, and even his daughter suffered from the same emotional pain.

But even in retirement Huxley was still a great thinker and contributor to secularism and humanist ethics. Writing and lecturing on *Evolution and Ethics,* he discussed at great length his ideas concerning the biological elements of our humanity and those that are culturally constructed. He clearly states that organized religion has not served and cannot serve to provide a moral or ethical structure for any humans to follow.

However, Huxley as a materialist concluded that we should make moral and ethical choices based on informed, unbiased, and conscious reason and not based on religious doctrine, which he viewed as culturally contrived.

When the world lost Charles Darwin in 1882, Thomas Henry Huxley served as one as the pallbearers at his funeral. There had always been a deep respect between the two men. Darwin certainly respected Huxley's scientific ambitions and his mind and had a vested interest in Huxley's brash methods and confrontational debate style. Huxley certainly understood and valued the time and attention Darwin put into his work, while feeling a bit angry at Darwin's lack of resolve to publicly speak about his ideas and liberate and educate the uninformed.

On June 29, 1895, Huxley himself died of a sudden heart attack after suffering for years from the dense smog of central London. Upon his death, he was laid to rest and celebrated by members of every economic class, by the leading scientists and politicians of the day, and by his bereaved and large family. Huxley is buried in East Finchley Cemetery in the Marylebone section of Central London.

The Family Huxley: T.H's Familial Legacy and Accomplishments

T.H. Huxley's children and grandchildren would lead prosperous careers and be part of future generations of British social, political, and intellectual life. Huxley's son Leonard, who was born in 1860, would marry and father scientist Julian, novelist Aldous, attorney David, and Noble Prizewinner Andrew. Following the same boy–then–two-girl pattern came Rachel and Henrietta Huxley. They too would marry and lead public and private lives. Finally, the two youngest children born in 1865 and 1866 respectively were Henry and Ethel, and each would also marry and live until the mid-twentieth century.

Henry would become a beloved and well respected doctor and Ethel would marry the artist John Collier in 1889. Collier's marriage to Ethel would be the second time he took wedding vows with a Huxley daughter. Collier was previously married to and then widowed with the loss of Ethel's late sister Marion, who died in 1887.

Many of T. H. Huxley's eight children were prodigious in their procreation as well. Son Leonard would have two wives and six children. Sadly, one of Leonard's sons, Noel, who was named for his late uncle who had died as a child, lived only to twenty-five and committed suicide in 1914. Leonard's only daughter, Margaret (1899), lived until 1981 and never married or had any children.

T. H.'s last-born son, Henry Huxley (b. 1865), was a prominent physician. He would marry wife Sophie and the two would in turn have five children. Henry's family would include two girls and three boys.

The elder Huxley's five daughters—Rachel, Jessie, Henrietta, Marian, and Ethel—would all marry. Rachel married twice and had five children. Jessie and her husband Fred had two children and several grandchildren. One of them was Sir Crispin Tickell, a much-lauded British diplomat and environmentalist.

Many of Leonard's and his sisters' extended broods of grandchildren and great-grandchildren are alive today and are living mainly in England and the United States, but their international legacy and current work within the arts, science, law, medicine, and politics (to name a few) can be felt around the globe.

T. H. Huxley's grandson and Leonard's son Julian Huxley was an outstanding biologist in his own right and one of the scientists who developed what is now called the modern synthesis of evolution in the early and mid-twentieth century. This synthesis pulled together Mendelian genetics, mutation, and natural selection. He had many other accomplishments, including being the first director-general of the United Nations Educational, Scientific and Cultural Organization (UNESCO) and first president of the British Humanist Association (now known as Humanists UK). He also chaired the founding meeting of the International Humanist and Ethical Union (IHEU) in 1952.

Even as T. H. Huxley's grandchildren would marry and disperse into the fabric of British, American, and European kinship lineages, connections to the Darwins would be everlasting. David Huxley, Leonard's son and T. H.'s grandson, had a daughter Angela who married George Darwin, Charles's great-grandson. Their 1964 wedding ceremony cemented in the bonds of matrimony a familial and multi-generational legacy between the Huxleys and Darwin-Wedgewoods.

CHAPTER 8

JOSEPH DALTON HOOKER

"I knew of this theory fifteen years ago. I was then entirely opposed to it; I argued against it again and again; but since then I have devoted myself unremittingly to natural history; in its pursuit I have travelled round the world. Facts in this science which before were inexplicable to me became one by one explained by this theory, and conviction has been thus gradually forced upon an unwilling convert."

Joseph Dalton Hooker

Briefly:

Joseph Dalton Hooker was a dear and long-time friend of Darwin. Hooker was also a naturalist and throughout his life was one of Darwin's most trusted confidants and editors. In 1858, standing in for both Darwin and Wallace, yet still managing to give priority to Darwin, he carefully made the world's first public presentation that described evolution by natural selection. He made it to the Linnean Society, the world's oldest active biological society. In late 1859, after the publication of *Origin*, Hooker published a book on botany that fully supported natural selection, becoming one of the first outspoken and respected scientists to adopt Darwin's grand theory. In 1860, he teamed with T. H. Huxley and famously debated evolution (and won) against Bishop Wilberforce and Robert FitzRoy at Oxford University.

If Huxley is known as "Darwin's Bulldog" then Hooker should be known as "Evolution's Midwife." Because of Hooker's friendship with Darwin, and his admiration for Darwin as a scientist, he wound up being in attendance at every stage of the presentation of Darwin's radical idea to the outside world.

To begin, he was the first scientist to whom Darwin ever confessed his radical idea. What's more, during the course of their lifetimes, Darwin and Hooker exchanged over 1,400 letters. Once Darwin had made that confession to Hooker, Hooker became an invaluable sounding board for the development of Darwin's theory.

Second, Hooker arranged the crucial meeting of the Linnean Society in 1858 when evolution by natural selection was first presented to the scientific world. He personally read both Darwin's and Wallace's work aloud at that meeting, and arranged for the publication of those works. Third, he argued bravely and formidably for Darwin's ideas at the infamous 1860 Oxford Evolution Debate. He continued to support both Darwin and evolution for the rest of his life.

Hooker's Birth, Contributions, and Life

At Darwin's home in Kent there are three portraits hanging over the mantle in his study where he did the majority of research for *Origin* and for all his later books. The first portrait is of Charles Lyell, the eminent geologist and scientist who Darwin admired greatly and whose book, *The Principles of Geology*, helped Darwin form his ideas regarding natural selection. The second portrait is of Josiah Wedgewood, Darwin's remote biological relative and, from a kinship perspective, his father-in-law.

The third and probably most important portrait is of his longtime friend, trusted confidant, and editor, J. D. Hooker. Hooker saw Darwin's mastery early on and encouraged Darwin's research and writing throughout his life. He was Darwin's loyal proponent and protector. Hooker helped Darwin build connections with naturalists and other scientists the world over. He also encouraged Darwin to expand the scientific boundaries of our understanding of who we are and how our species relates to the rest of the natural world.

Hooker came from an esteemed line of intellectuals and naturalists. However, though this description might give the impression that he was a bored member of the aristocracy or just a man in waiting, in reality nothing could be farther from the truth. Hooker was a naturalist who earned his reputation as a discoverer and leader of numerous international expeditions, sometimes quite dangerous, all in an effort to document the biodiversity of the world.

Just as in Darwin's clan, wealth and privilege provided a means of access for the young Joseph to explore the world. Later on, just as with Darwin, an experienced Hooker would become a leader in the field of biology and natural history.

Hooker's family had for decades contributed to British society and was deeply involved with science, exploration, and the biological world. Hooker's father, William Jackson Hooker, was the director of the Royal Botanic Gardens at Kew for many decades, a position that Joseph himself would apprentice for and eventually obtain.[131] During Joseph Hooker's time in the leadership, Kew Gardens were expanded even beyond what his father had done.

Early Life and Education

Hooker was born on June 30, 1817, in Halesworth, England. Halesworth is near Britain's east coast and is situated about 110 miles (about 177 km) from London. The hamlet at the beginning of the nineteenth century was sparsely populated and Hooker's early life was spent near home and then, as a small child, in Glasgow, Scotland. Hooker developed his interest in botany at a young age and studied the exploits of early explorers, combining his interest in tales of exotic voyages with his father's botany lectures at Glasgow University.

These two fundamental elements, situated between travel and botany, would form the core two pillars of his later experience and would shape his interests and motivations across his lifetime of study, writing, and science advocacy. But many of his exploits and accolades would come later. Hooker at twenty-two, always the serious student, would obtain his medical degree in 1839 from Glasgow University.

Hooker's Early Career—
Antarctica, Evolution, and the British Geological Survey

For all his advantages, in his early career Hooker had a hard time finding secure employment. But as with Huxley, his education in medicine enabled him to join the Naval Medical Service[132] and thereby become part of a scientific expedition. He would make many voyages throughout his life, in one capacity or another. In 1839, he set sail on a four-year research trip to the Antarctic.

Hooker was the youngest member of this voyage and went in the role of assistant surgeon. In addition to acting in a medical capacity, Hooker assisted the ship's surgeon in making zoological and geological collections. He also made botanical collections, as Huxley did on his voyage, gathering specimens right out of a net that was dragged behind the ship, thereby obtaining samples of algae and other sea life.

The voyage was the last major voyage of exploration done entirely under sail. The two sailing ships were cheerfully named *Erebus* and *Terror*. Hooker had his berth on the *Erebus*. The Ross Ice Shelf in Antarctica is named for the commander of that expedition, James Clark Ross.

On his trip Hooker took along a copy of Darwin's *Journal of Researches*, the book now known as the *Voyage of the Beagle*. It was a gift from Charles Lyell, the father of the geologist who had so influenced Charles Darwin. In addition, before embarking, Hooker had the good luck to briefly meet Darwin, courtesy of a mutual friend. Hooker had already read some of Darwin's travelogues and been inspired by them. During the voyage, Hooker read the *Journal of Researches* and was impressed by Darwin's work as a naturalist.

The twenty thousand mile round-trip voyage included travel though South America and the Falkland Islands, around the Cape of Good Hope, stops in New Zealand, and finally several months in Antarctica itself, mapping the land, collecting specimens, and maintaining records of new species and finds. All these items would be fully classified and cataloged upon Hooker's return to England in 1843.

Hooker was thrilled by the idea of being able to explore where no botanist had gone before, and where none would probably ever go again! He wrote:

> "No future botanist will probably ever visit the countries whither I am going, and that is a great attraction."[133] [J. D. Hooker in a letter to his father before the *Erebus* set sail, 3 February 1840]

Hooker was simultaneously bold in his researches and hard working in publishing his scientific results. The Antarctic voyage made Hooker a name as an intrepid botanical explorer.

Hooker and Darwin started corresponding in 1843. Not long after Hooker got back from the Antarctic, Darwin wrote to him and asked if he would be interested in classifying the botanical specimens that he had brought back from the Galapagos. Somewhat surprisingly, it was only a few months later, in 1844, that Darwin let Hooker in on his radical idea, which we now call evolution by natural selection:

> I am almost convinced, (quite contrary to the opinion I started with) that species are not (it is like confessing a murder) immutable.... I think that I have found out (here's presumption!) the simple way by which species become exquisitely adapted to various ends.[134]

In reality, Darwin had briefly sketched out his ideas about natural selection in 1839. These he expanded into an essay in 1842 that was thirty-five pages long. In 1844 he expanded it still further to a 230 page "essay," which others might call an unpublished book. In fact, Darwin realized that it might make an adequate book about natural selection if he were to die early. He asked his wife Emma to see to it that the work got edited and published, and suggested Hooker as one possible editor.

Darwin offered his 1844 essay to Hooker to read in 1844, though Hooker did not actually do so until 1847.

Meanwhile, Joseph Hooker had to make a living. Upon his return to England, with a dynastic sense of accomplishment along with a reputation as a promising young scientist, Hooker's next formal expedition was closer to home. He enlisted as a principle researcher in the British Geological Survey. Hooker had become an expert in British paleontology, learning from the early and later works of Henslow and Lyell, much as Darwin did during the course of his research on the *Beagle* and at Down House.

From February 1846 through October 1847, Hooker worked with the fossil plants found in coal-containing rocks. This was a very energetic era in the history of the survey because it had been tasked with producing a geological map of Great Britain in 1845. Thus, in the summer of 1846, Hooker investigated the coal fields of Bristol, Somerset, and South Wales. That fall he went back to London and started cataloging the specimens. Thus, as with Huxley, the then-thriving British interest in fossils gave him employment in a geological capacity. As usual, Hooker was energetic in working and publishing his results. However, by the time they were published he was off on his next adventure.

Voyages to India and the Himalayas, Palestine, and Morocco

In 1847, Hooker's father soon to become the director of the Royal Kew Gardens, recommended that his son take part in an expedition to the Himalayas to collect botanical specimens. While Darwin's father grudgingly protested but then paid for his son's exploits, Hooker's father compelled his son to explore with great eagerness and earnestness in order to make Kew Gardens *the* world-class botanical wonder of the nineteenth century. Such activism to create a collection with size, depth, and diversity would be a source of pride for the Hooker family throughout many generations, even up to today.

In November 1847, Hooker and a large group of assistants, surveyors, and scientists left the United Kingdom for India.[135] This group would serve as the intellectual and diplomatic brain trust for what would become an expedition lasting four years. More years again spent in discovery, with the mapping of land and the hunt to identify and catalog hundreds of new plant species. By early 1848, Hooker and his team would set up their base camp in Darjeeling and, taking turns going out in a foreign and sometimes dangerous land, map and collect specimens. If the elements could at times be harmful, the indigenous people could be downright dangerous.

Perhaps one of the most striking stories while on the Himalayan expedition, as it traveled through Tibet, was the capture and confinement of both Hooker and another Briton, Archibald Campbell, by the Dewan of Sikkim.[136] It took a British diplomatic team to save them, along with Hooker advocating for himself and his companion to not be tortured and left for dead as unwanted and unholy intruders onto the Dewan's native lands. Once set free, Hooker, on the promise of remaining a respectful distance for the rest of the expedition, spent more time reprioritizing and focusing on writing up his experiences, specimens, geographical mapping, and observations rather than going out into the field. His work culminated in part with the later publication of *Himalayan Journals* published in 1854 and dedicated to none other than his best friend, Charles Darwin.

Hooker would mount three other major botanical exhibitions during his long life. He traveled to Palestine in 1860, observing the biodiversity of the region and making strong inferences on the plant relationships across the Sinai. Then in 1871, Hooker traveled to Morocco to collect specimens for Kew. Finally, in 1877, as colleague and friend to Asa Gray, he explored the American West, including Utah and Colorado, which led to the discovery of dozens of new plant species that were shared equally between Gray in the United States and Hooker in the United Kingdom. (The western expedition will be discussed separately, later in this chapter.)

Hooker's research also convinced him that the cedar species in Lebanon, *Cedrus libani*, and the Moroccan cedar species *Cedrus atlantica*, are simply varieties of the same species. He further came to believe that the climate had previously been much colder, and that there had perhaps once been a continuous forest at lower levels. As the climate warmed, these cold-weather species retreated up into the mountains into isolated high pockets.[137]

Isolated populations like these ancient cedars, separated from other populations like them by inhospitable desert, can be considered, biologically, to be similar to island populations. In the case of islands, the inhospitable sea separates the populations, and in the case of high mountains surrounded by deserts, the low, arid, and inhospitable desert also serves to isolate high mountain populations of plants from one another. One extremely important fact to remember, however, is that in the case of the Lebanon cedars, the ecological community was already well established when the Lebanese portion of the ecosystem became separated from the Moroccan portion. Once the communities become isolated from one another, no more gene exchange can take place between members of the different isolated populations. From that point onward, the species in the two populations will evolve separately from one another. Eventually this may lead to different species being formed in each of the original populations. Today this is known as allopatric speciation. In the case of the cedars in Lebanon and Morocco, Hooker thought that the cedars in these two separated places were not yet separate species, though they were slightly different from one another. But other people thought that they had reached the point of being different species.

Hooker's research in the Syria of the day convinced him that Syria's vegetation could be divided into three regions: the western or coastal region; the valley of Jordan, the Dead Sea and the areas surrounding Damascus; and the middle and mountainous regions. In the latter, Hooker found that oak trees predominate in the areas below three thousand feet. He counted the rings on one fallen oak branch and found it to be about seven hundred years old! This type of botanical mapping was useful both scientifically and economically.

Empire and Botany

Why did British botanists like Hooker collect plants from all over the world? Why were these specimens studied, and often planted, at Kew Gardens? Why were maps of worldwide plant distributions considered so important? Why did the British government sometimes support such expensive studies? The reason had a bit to do with pure science, and much more to do with the economics of the British Empire, whose wealth was largely based on plants!

To begin with, all that plant population mapping resulted in maps. Maps were useful to the British government both economically and militarily. Some of the maps that Hooker made were used in military campaigns. However, the British also needed to know what their empire contained. Plant materials like cotton, rubber, indigo, plants used for dyes, timber, sugar, spices, tea, and even rhododendrons were valuable commodities. Hooker's bringing back many additional different types of rhododendrons from his Himalayan expedition, and writing a book about them, is said to have ignited a rhododendron craze among English gardeners.

For Kew, Hooker not only collected plant specimens himself, he also obtained plant specimens from a network of collectors who sent him both plants and seeds. These could then be classified if necessary and grown at Kew if the species was economically valuable. The plant maps could then be consulted to find other parts of the world that had similar growing conditions, and those economically beneficial crops could then be grown in areas controlled by the British crown. For example, Kew Gardens obtained seeds of the rubber tree, *Hevea brasiliensis*. These were planted and raised at Kew. Young rubber plants from Kew were next sent to Ceylon and Malaya (as those places were called at the time), and this was the beginning of the rubber industry in Asia. The rubber thus obtained allowed Britain to not only make money but to also retain its dominance in cable telegraphy. This in turn helped to keep the empire connected and running, and helped with military communications.

One of the ongoing lessons in the history of science is how basic scientific work, just data collection and careful descriptions, often produces unexpected results. For example, practical research for empire's sake, as Hooker did, resulted in the basic data collection on plant distributions being done. Once done, patterns emerged, and fundamental questions (and sometimes answers) were exposed by the data. Botanists Hooker and Gray were both working on gathering data about plant distributions, and both wound up with similar questions and problems being exposed by their work.

These questions were then answered by Darwin's theory of evolution by natural selection, which is why both Hooker and Gray became its strong proponents. Similar sorts of questions about the distributions of animal populations led Wallace to become the co-discoverer of evolution by natural selection. Major advances in human understanding, such as our understanding of evolution, have then in turn shaped our knowledge and well-being. In fact, our well-being would be further enhanced if evolution's predictions were taken seriously and acted upon appropriately in the present day. For instance, the evolution of antibiotic resistant bacteria is entirely predicted by evolutionary theory. Unfortunately, evolutionary theory was ignored and, as a result, people are now dying of diseases that we used to be able to cure.

Before evolution could inform every aspect of biology, however, it first had to be introduced to the wider world.

Hooker's Role in Introducing Evolution to the World

Although most people first think of Huxley in the Wilberforce debate of 1860, Hooker was the pivotal individual in the original public presentation of evolution by natural selection at that germinal meeting of the Linnean Society on July 1, 1858. It was Hooker who read aloud both an extract of Darwin's 1844 essay, a letter from Darwin describing natural selection to the botanist Asa Gray, and Wallace's independently derived essay on the same subject.[138]

Wallace, meanwhile, was pleased with the recognition that his work received, and was happy that he and Darwin had both been credited with the discovery. Hooker and the geologist Charles Lyell were the only two scientists who had read Darwin's 1844 essay, and were also the only two people besides Darwin to read the essay that Wallace had sent to Darwin. Darwin had forwarded Wallace's essay to Lyell, who had then forwarded it to Hooker. Between them, they decided that Darwin *and* Wallace's writings on evolution would be read, as near to simultaneously as possible, at that meeting. Hooker and Lyell wanted Darwin to have priority, but had to be fair to Wallace as well. Darwin himself was anguished about being pushy about who came up with the idea of natural selection first, but he wasn't the one who organized the meeting. That was done by Hooker and Lyell.[139]

In a letter to Hooker, Wallace wrote:

> Allow me in the first place to thank yourself and Sir Charles Lyell for your kind offices on this occasion, and to assure you of the gratification afforded me for both the course you have pursued, and the favorable opinion of my essay which you so kindly expressed. I can't but consider myself a favored party in this matter, because it has hitherto been too much the practice in cases of this sort to impute all merit to the first discoverer of a new fact or a new theory, and little or none to any other party who may, quite independently have arrived at the same result a few years or a few hours later.[140]

Wallace went on to say,

> it would have caused me much pain and regret had [his] generosity led him to make public my paper unaccompanied by his own.[141]

Thus Hooker and Lyell managed to navigate the delicate task of getting evolution by natural selection out into the public scientific sphere, crediting its two discoverers simultaneously while still giving Darwin priority, all while keeping both co-discoverers satisfied with the result.

Hooker was in attendance at the famous Saturday meeting of the British Association, now known as the Oxford Evolution Debate. But he was there reluctantly. He went to Oxford during all the days of the British Association's meeting but really didn't want to put up with the nonsense that was going on.

He wrote to Darwin on July 2, 1860:

> I came here on Thursday afternoon & immediately fell into a lengthened reverie: without you & my wife I was as dull as ditch water & crept about the once familiar streets feeling like a fish out of water—I swore I would not go near a Section & did not for two days—but amused myself with the Colleges buildings & alternate sleeps in the sleepy gardens & rejoiced in my indolence. Huxley & Owen had had a furious battle over Darwins absent body at Section D.,[3] before my arrival,—of which more anon. H. was triumphant— You & your book forthwith became the topics of the day, & I d—d the days & double d—d the topics too, & like a craven felt bored out of my life by being woke out of my reveries to become referee on Natural Selection &c &c &c—[142]

One wonders what was going through Hooker's mind. Why did he go to Oxford at all during that week if not to attend the meeting of the British Association? One gets the sense that he didn't like the circus atmosphere but couldn't stay away. He undoubtedly wanted to hear how the meeting was going, and decide on a daily or even hourly basis whether or not he wanted to get involved with the goings-on.

But Saturday was different. Everybody knew that the Bishop of Oxford, Samuel Wilberforce (referred to here as Sam Oxon) was going to speak. Darwin's apostles had no reason to trust this talented and slippery speaker. Wilberforce wasn't known as "Soapy Sam" for nothing. So Hooker decided to go to the Saturday session.

In that same letter he reported this from that crowded, famous Saturday encounter:

> The meeting was so large that they had adjourned to the Library which was crammed with between 700 & 1000 people, for all the world was there to hear Sam Oxon—Well Sam Oxon got up & spouted for half an hour with inimitable spirit uglyness & emptyness & unfairness, I saw he was coached up by Owen & knew nothing & he said not a syllable but what was in the Reviews—[143]

Remember Richard Owen? Well, Hooker figured out who had coached old Soapy Sam the clergyman who knew nothing about biology. The meeting threatened to get out of hand. Lady Brewster fainted. Huxley had done his best, had gotten in his dig, but didn't have a voice that carried. Eventually Hooker became so infuriated that he decided to speak. He asked John Henslow, the man running the meeting, to let him address the audience.

He went on:

> The battle waxed hot. Lady Brewster fainted, the excitement increased as others spoke—my blood boiled, I felt myself a dastard; now I saw my advantage—I swore to myself I would smite that Amalekite Sam hip & thigh if my heart jumped out of my mouth & I handed my name up to the President (Henslow) as ready to throw down the gauntlet—[144]

And he described his (and Darwin's) victory:

> it moreover became necessary for each speaker to mount the platform & so there I was cocked up with Sam at my right elbow, & there & then I smashed him amid rounds of applause—I hit him in the wind at the first shot in 10 words taken from his own ugly mouth—& then proceeded to demonstrate in as few more 1 that he could never have read your book & 2 that he was absolutely ignorant of the rudiments of Bot. Science—I said a few more on the subject of my own experience, & conversion & wound up with a very few observations on the relative position of the old & new hypotheses, & with some words of caution to the audience—Sam was shut up—had not one word to say in reply & the meeting *was dissolved forthwith* leaving you master of the field after 4 hours battle.[145]

So after helping Darwin at every stage in the presentation of his theory, Hooker got to deliver the literal last word on it at the Oxford Evolution Debate. Emotion and unreason had not been allowed to carry the day.

It should be noted that Hooker recognized that Wilberforce had been coached by Richard Owen, and it should be remembered that botanical evidence played an important role in providing evidence for evolution. A specialist in zoology such as Owen would most likely have not understood botanical evidence, just as Hooker claims in this letter. Since Wilberforce himself knew no biology, or even any science of any kind, and Owen was knowledgeable only in zoology, this made Wilberforce's arguments vulnerable to attacks by a massively knowledgeable botanist such as Hooker, who knew his botanical evidence, and knew it well.

Hooker's Later Career—Kew

Although Joseph Hooker's botanical career took him all over the world, he simultaneously wound up with a career at the Royal Botanic Gardens at Kew. In 1855, when Hooker was thirty-eight years old, he took a position as assistant director of the Botanical Park.[146] From then on, with his father as his champion, Hooker's fate as one of England's top botanists and esteemed naturalists was sealed.

But Hooker's time as Director of Kew, which began in 1865, would not be as serene and calm as that of his father. Controversy would surround Hooker for two reasons, each out of his control and each caused by men who felt they and their careers were superior to Hooker's skills and position. The first "troublemaker" for Hooker was Sir Richard Owen, the Director of the British Natural History Museum, who felt that Kew should be managed by the museum and that Hooker, rather than being independent, should report directly to him.

In Owen's mind, the real blasphemy of Hooker was his defense of Darwin and natural selection. Owen was a proponent of a kind of "intelligent design" and while he received a complimentary copy of *Origin* directly from Darwin, Owen could never bring himself to see the natural world as acting without a divine hand. Owen would work to professionally discredit Hooker at every turn because of his evolution advocacy and because of the last strike, which was Hooker's involvement with the X Club. The X Club represented a space where a new intellectual order was working to overturn long-held ideas of science, which Owen held onto with vigor.

The other person who worked to push Hooker out of Kew was Acton Smee Ayrton, a British politician, ignorant of science and the work of botany, who saw Hooker's work and Kew Gardens as an elite dalliance that was simply not needed. Ayrton became the governmental overseer of Kew, crippling its budget and removing Hooker from making any hiring decisions. He worked on the inside with his connections in Parliament to defund Kew. In fact, Ayrton commissioned a special report on Kew, which claimed Hooker had mismanaged funds and damaged the park's trees. Hooker never saw the report although it was shared within the government. And while this may seem coincidental, the expose's main writer was the one and only Richard Owen.[147]

So there was clear collusion and bias in writing about both Hooker and Kew. And it would take outside advocacy by many scientists (including that of the apostles and Charles Darwin) who would write to the government in support of Hooker. In fact, once these men of prominence were mobilized, it did not take long for Hooker's reputation to be reinstated and his work at Kew to be fully supported by the government with deep respect and honor.

Voyage to North America, 1877

In 1877 Hooker made a trip to the western United States with another of Darwin's apostles, the Harvard botanist Asa Gray. The journey centered on the Rocky Mountains, and both Gray and Hooker hiked and climbed their way over many tall western peaks. Why did two now-elderly botanists make this arduous trip? The reason for this expedition, besides pure collecting, deserves some explanation.

Why Botanical Evidence is So Important to the Study and Acceptance of Evolution

Studying the geography of plant distribution, known as plant biogeography, was work that could be done by scientists in Darwin's time without having to wait, and hope, for more fossil discoveries. Worldwide plant communities were (and are) living laboratories where the predictions made by evolution by natural selection could be tested. For this reason, plant distribution around the world played an enormously important role in providing evidence for evolution by natural selection.

Hooker's ideas grew in favor of natural selection as his years of travel, research, and observation developed his understanding of the core elements and ideas of evolution as postulated by Darwin. As a botanist, Hooker knew that if individual plant species evolved, then they would have a single point of origin, and then they would spread from there. As they spread, they would adapt to the challenges of their new regions. As they adapted, they would develop new characteristics, which would demonstrate that they had changed because of the environmental pressures provided by their new surroundings.

If, on the other hand, species were created by God, then if the same species of plant was found in different parts of the world, then God must have created it separately each time, in each region.

The outcomes of these two different modes of dispersal would look different, and there were differences that careful scientists could observe and measure. Darwin himself, though known for his work on animals, was also a world-class botanist. In some respects the two botanist-apostles on this trip were acting in part as proxies for Darwin as they did their research.

Hooker collected over a thousand specimens on his U.S. trip. What Hooker and Gray found in part was that the vegetation in the high Rockies was more similar to plant communities in the Altai mountain region of central and East Asia than they were to plant communities either on the East Coast or West Coast of the United States.

Hooker and Gray published a paper in 1882 about this research in Volume Six of *Hayden's Bulletin of the United States Geological and Geographical Survey of the Territories*. Its title was "The Vegetation of the Rocky Mountain Region and a Comparison with that of other Parts of the World."

What they found provided evidence for the idea that, prior to the last ice age, there had been a cold but relatively milder climate in the arctic regions of the world, where cold-hardy plant communities existed. As the ice age started and the ice sheets formed, these cold-tolerant but not ice-sheet-tolerant plant species had been driven south in front of the ice. As this happened, the plants that had been in those more southerly areas were driven out of their ranges by the severe change in temperature and the competition from the cold-weather plants from the arctic. In mountain ranges like the Rockies, these plant communities lived on the southerly edge of the ice fields. Then, as the ice receded, the cold-loving plant communities followed the ice northward and wound up remaining in the cold high-altitude upper reaches of high mountains in various places around the world, including the Rockies.

Their data therefore supported the idea that similar plant species were not created separately (and divinely) in different parts of the world but rather had a common community of origin from which they were pushed to various points on the globe by changes in the climate of the planet. This is how true Arctic plants wound up in high mountain ecosystems in parts of the world that are far removed from one another.

Going Deeper: Hooker's Friendship with Darwin and How His Life's Accomplishments Deserve to Be Told

Hooker's writings are wide as they were deep. They mainly focused on the one topic that totally consumed Hooker's main academic and naturalist interest: to find and later classify species of plant life previously unknown to humanity. So his writings mainly focused on the discoveries he made during his expeditions. Beginning in 1844 with the publication of his findings in Antarctica (*Botany of the Antarctic Voyage of the Ships Erebus and Terror*) to his later expeditions (*The Flora of the British Isles*), as well as other works including *The Rhododendrons of SikkeHimalias* to the *Handbook of New Zealand Flora*. All told Hooker would publish almost twenty books covering the world's ecology and flora. His numerous other writings, reviews, and papers show an almost inexhaustible interest in almost every corner of the natural world.

It is clear that the deep friendship and lifelong partnership of Hooker and Darwin made a lasting mark on British science. It was Hooker who Darwin trusted with his manuscripts from *Voyage of the Beagle* through *Origin* and the *Descent of Man*, along with Darwin's numerous other books and scientific papers, long before each was published. It was Hooker who spoke most eloquently on Darwin's behalf in 1860 as Huxley's second and as a more adroit defender of Darwin in that same Huxley-Wilberforce debate. It was Hooker who wrote reviews of *Origin* which appeared in scientific journals in England, and it was Hooker who more importantly served as Darwin's intermediary with the British and international scientific community throughout their lives.

During the apostle's time, many people did not comprehend the evidence for evolution. After all, multicellular species do not usually evolve in front of people during their lifetimes. What's more, when *Origin* was first published, there had not yet been any fossils collected that showed a clear transition with one species in the act of becoming another.

As luck would have it, the first shatteringly convincing transitional fossil was *Archaeopteryx*, which was first described in 1861. This fossil, found in Germany, showed an animal that had characteristics of both dinosaurs and birds, including wings and feathers that are characteristic of birds, and teeth that are characteristic of dinosaurs.

Although Darwin discussed *Archaeopteryx* in later editions of *Origin*, more evidence needed to be gathered in order to convince a skeptical public of the truth of evolution by natural selection. In addition, at the time, people also did not understand modern genetics, which in later years would provide more evidence still for evolution.

However, as discussed earlier, in terms of science, the *botanical* evidence for evolution could be gathered during Darwin's lifetime. And it was—provided by Hooker and Gray. This is because plant specimens could be obtained by the thousands. For every one lucky fossil find obtained by fossil hunters, botanists could look at a thousand plants. The painstaking work done by botanists helped provide overwhelming evidence for Darwin's ideas.

Hooker was a dearly trusted family friend of both Darwin and his wife Emma, something that few of the other apostles could rightly claim, although they certainly had cordial relations with the entire Darwin clan. Perhaps it was because both Darwin and Hooker spoke the same intellectual language—that of science, that of social caste, that of education and of interest in the natural world. These were elements that would link these good friends while they were alive and keep them linked, as monuments to reason and modernity, in their passing.

Family Life

Hooker's personal and family life was deeply complicated. In many ways he was a man at the helm of what can best be described as a modern blended family. In 1851 Hooker married Frances Henslow. If the last name reads familiar it's because Frances was the daughter of John Henslow, the famous British botanist and geologist, and mentor to none other than Charles Darwin. Joseph and Frances would have seven children, all but one living well into the twentieth century. Frances was key to Hooker's early success as a naturalist as she was frequently called upon to translate much of his research and writings from and into French.

The Hooker children all were genetically blessed with the appropriate genes for long life, as was Hooker himself. Sadly though, as if history is nothing more than a reflection of his and our times, it is clear that the loss of his daughter, Mary Elizabeth Hooker, at the age of six would be a true emotional blow to Hooker for the rest of his life. He saw Mary as his gem and felt her loss much as Darwin felt the loss of his Anne at the age of nine in 1854.[148]

It appears that Darwin and his apostles were united in both science and in the loss of favored children. And while death was kind to most of Hooker's other children it would stalk his family still, taking his wife Frances from the family in 1874 when she was forty-nine. Two years later, while Hooker was busy working at the Royal Gardens at Kew, a widower with a large family, he would marry Hyacinth Jardine, who was twenty-five years his junior. That marriage would last until Hooker's own death and it would also produce two more sons. Hooker was still producing offspring well into his fifties and his last child was born when he was sixty-eight years young!

All told Joseph Hooker would have nine children with two wives over the course of thirty-two years. His first was born in 1853 and his last in 1885. Most of his children would marry and the Hooker family tree is extensive with grandchildren, great- grandchildren, and great-great-grandchildren as living descendants today.

In fact, one of the authors of this book had the opportunity to meet with and interview Isobol Moses in 2016, the daughter of Reginald Hawthorn Hooker (1867–1944) and great-granddaughter of Joseph Hooker. At 70, Isobol was spry and movement oriented. She recounted with glee the family lore surrounding her great-grandfather. She expounded on the personal and psychic pain caused by the Owen witch hunt and the toll it took on Joseph Hooker, who at times would seem broken and moody to his family as a result of all that pressure.

But as a true Hooker, she said both she and great-granddad were optimists. She also recounted his happier times at Kew Gardens, the pride he felt in his work and his earlier explorations. Ms. Moses explained that Joseph Hooker always seemed "on fire" and that he never lost his passion for his work or for his family. She surmised that he was a tough man and lived an extraordinary long life.

She also felt that those genes were passed down to her and her kin as well. The author that met her can never forget how at seventy years young, while we were walking together, she leapt off a two foot high stone wall in the church cemetery where Hooker is buried. Landing gracefully on the ground below like an Olympian, she turned and smiled as he hesitated, found a more conducive path, and caught up to her. All the while she kept walking and talking about Joseph Hooker and the Hooker family tree.

Joseph Hooker died in 1911 after a brief illness at the ripe age of ninety-four. Because of his close connection to Charles Darwin, Hooker was offered a resting place near Darwin's at Westminster Abbey. However, those were not his wishes and, instead, he was buried close to the institution that would define his life's work (and that of his family) at the Royal Kew Gardens. Joseph Dalton Hooker is buried just meters away from Kew at St. Anne's Church, on a steppe of land not far from the beauty he helped create during his time as director.

CHAPTER 9

ASA GRAY

"Perhaps if zoologists would contemplate the wide variations presented by many plants of indubitably one and the same species, and the still wider diversities of long cultivated races from an original stock, they would find more than one instructive parallel to the case of the longest domesticated of all species, man."

Asa Gray

Briefly:

Asa Gray was born in the United States. He was a botanist and early colleague and collaborator of Darwin. Gray spent much time in England and, along with Hooker, did field research on Darwin's behalf. Back in the United States, Gray advocated for *Origin* and actually negotiated its first American printing as well as royalty rights for Darwin. Gray also wrote *Darwiniana*, a collection of essays written by Gray in support of natural selection and evolution. Gray was also a strong supporter and advocate for minimizing the opposing viewpoints between religion and science and worked to harmonize the distinctly different ways in which each viewed the world and humanity's place in nature.

Gray's Education and Life

The story of Asa Gray is a complex one. All great men of science have their trials and tribulations. Certainly for much of his life, Gray attempted to reach a parity with evolution and faith, stitching together the stark differences between science and religion. It was Gray who, as Darwin's chief champion in the United States, sought to bring together natural selection and faith as examples of healthy coexistence between "spiritual creative forces" and Darwin's grand theory. But as Darwin rightly knew, his grand theory did not need any help from the metaphysical world to be held as actual scientific truth.

Early Education

Gray was born in 1810, the first of eight children to his father, Moses, who was a tanner by profession, and his mother, Roxanna. They lived in upstate New York in the town of Sauquoit, settled and administered by Presbyterians who had earlier fled eighteenth century English religious persecution.[149] Gray's religious education started at the age of three. His father and grandfather, both named Moses, surely felt that it was their duty to bring little Asa and his siblings up in a strict world where the "word of God" meant something beyond the physical world. But while Gray would spend most of his early years learning the Bible at several different conservatively religious schools, his family neither stopped nor hastened the young Gray's interests in the natural world.

Gray attended a nearby grammar school, then went to Fairfield Academy in Fairfield, New York. Like Darwin and all the apostles except for Wallace, Gray also studied medicine, graduating from Fairfield Medical School in 1831. In his earliest life as a student and far from being the eminent biologist and botanist he would eventually become, it was Gray's interest in medicine that eventually brought the young scholar to medical school.

A man of his times in upstate New York, Gray's early years and intellect led to career opportunities that were similar to Darwin's. As a young man of promise, Gray's further education could lead to three distinct career paths, if his family could afford the schooling. Typically, he could become a physician, an attorney, or he could join the clergy. Since he was a young man with a scientific bent, it is not surprising that he chose to study medicine. Yet even as he studied to become a doctor, Gray was deeply curious about the natural world.

His interest in botany appears to have been piqued by an article in *Brewster's Edinburgh Encyclopaedia*, which he read in the winter of 1827-28. Studying plants, it should be noted, was still very much a part of a medical education in those days. Also, a person of a general scientific bent often wound up being called upon to work in a number of different fields, and such was the case with Asa Gray.

Early Career

Gray's first real mentor was Dr. James Hedley. Hedley would be a friend, contributor, and agitator for Gray's career for decades. For the young Gray who was barely old enough (in our time) to vote, Hedley's impact on his ideas would also last a lifetime. It was Hedley who further pushed on the nascent Gray's interest in minerals and the plant life surrounding his home. To Hedley, the natural world was a reflection of God's handiwork and he often would propose to Gray that "nature evidenced the work of God."

By 1828 Gray was fully involved with his medical studies but he was also immensely drawn to cataloging and classifying the natural world around him. Later in life, thanks to his interests and travels, it would be Gray who would document the diversity of plant life in the United States. At the age of twenty-one in 1831, Gray was awarded his degree of Doctor of Medicine. What may seem like a young age to be graduating medical school today was indeed in Gray's time not atypical. A compression of schooling often culminated in much different outcomes for American and European men (it was mostly men) who could be nurtured by mentors or within institutions. Thus they began working in their careers much younger than what is expected by today's standards and training.

Originally Gray, like most medical students, was happy to complete his formal studies. For a brief time he thought he might achieve a prosperous career as a professional man of medicine. However, such interest would never come to pass and he was never fully committed to the medical arts and sciences. His heart just was not into medicine even though he saw it as a noble field and one that certainly made his religious and socially conservative parents proud. Instead, his interests had been won over by the study of botany.

After graduating from medical school, Gray taught chemistry, mineralogy, and botany at a local high school, gave public lectures on botany at the medical school, and later gave a course on mineralogy and botany at Hamilton College.[150] He continued to be an active botanist, collecting specimens for his herbarium while in medical school, using money from his post-graduation jobs to make botanical excursions, and corresponding widely with fellow botanists. In fact, Gray actually never practiced medicine, the field in which he first studied and was degreed.

Perhaps field botany was indeed his first true academic love and one that, as he grew older and more experienced in the field, he could no longer ignore. While Gray would have made an excellent doctor, saving and healing countless lives, our true loss would have been not having him spend the next decades of his life studying the flora and fauna around us. In the end, Gray had hoped to make a career studying plants and while not formally trained as a botanist, his career goal was to meld medicine and botany, since understanding the natural healing power of plants was a sure fit in his overall area of medical training and burgeoning expertise.

In perhaps another interesting parallel between the two men, both Gray and Darwin would reject their medical training for a chance to unlock and better understand nature. But while Darwin would leave his medical studies it was Gray, who was less wealthy and thus could not opt for other choices as a young man, who completed his training to become a physician.

In the United States during the mid-nineteenth century, medicine was seeing itself recast in ways which professionalized and standardized both the field and the science behind how to heal the human body. This meant recognizing the contributions of not only the practice of medicine but also the allied fields that often led to better techniques and greater discoveries in the medical arts. As with most happenstance, this was also a good time for now Dr. Gray, with his professional connections, to try to recast himself as a naturalist and botanist.

Because of the death of Dr. Fay Edgerton, an eminent botanist, Gray was able to in essence luck himself into a teaching and research position at the Utica, New York, Gymnasium. It is here where Gray would, through numerous field expeditions, focus mainly on gathering and classifying hundreds if not thousands of plant and animal species.

Gray's academic appointment at the Utica Gymnasium legitimized his interests and allowed him to work all over the tristate area collecting plant species. What a glorious and exciting time it must have been for him and his field assistants who captured and studied never before acknowledged or understood plant species in New York, New Jersey, and Pennsylvania.

By 1834 Gray's descriptive work had become acknowledged and well accepted. At the Lyceum, one of New York's strongest academic societies of the day, his accomplishments had been lauded. This acceptance essentially provided the recognition which Gray had earned through his fieldwork, observation, and classifications. It would be momentous, as it would also put him into contact with biologists and naturalists from around the world.

Another of Gray's early mentors at this time was Dr. John Torrey, a colleague who would eventually connect him to another of Darwin's apostles, Dr. Joseph Hooker. But however exciting this time was intellectually for him, it was also very stressful because he was financially insecure and had been at times, even with the Utica position, living by his wits. This lack of financial solvency depressed the young Gray and also made him quite despondent at times. His name recognition in scientific circles was growing just as his anxiety about his personal future was cast into doubt by his inability to secure money.

It was Torrey, a professor of Chemistry and Botany at the New York College of Physicians and Surgeons, who hired Gray in 1833 to be his assistant. For financial reasons the position did not last, but in 1836, with help from Torrey, Gray was appointed to be the curator and librarian of the New York Lyceum of Natural History. In addition to publishing original scientific papers, he also published his first textbook, *Elements of Botany*, in 1836.

But as celebrated as Gray was becoming within the field of botany he was still financially insecure. Many famous and brilliant men and women are faced with the same set of financial quandaries—talented in their field but not able to make ends meet or just unable to manage money. We can remember Mozart as one of the most important classical composers of his and our time but as someone who was buried in a pauper's grave. So too was the fate of Edgar Allen Poe, Herman Melville, and Oscar Wilde, who each died penniless. In modern times, entertainers such as Judy Garland and Sammy Davis, Jr. each died owing millions to creditors but were indeed world famous. So too was the fate of inventor Nikola Tesla, who lived in poverty and died penniless although his patents helped give us our modernity. Placed into context, while Gray lived before and after some of the great minds and celebrities mentioned, the very real threat of poverty did not escape his psyche.

As a result of his insecurities about his future in science and his finances, Gray returned to some familiar metaphysical ground. He joined the Bleeker Street Church in 1835 for spiritual guidance and emotional comfort, two things he could not find within his secular and scientific life.[151] It was here at the church that Gray convinced himself that science and faith were indeed compatible, a belief he held to his last day.

Gray also taught Bible classes and was enthusiastic at trying to link the magisteria of science and faith. This bifurcation could not be clearer than in 1836, when Gray published *Elements of Botany* to much fanfare and public acceptance. In fact, one could certainly call him a popularizer of science for the masses. For many years, he published a series of botanical treatises meant for educational purposes, including a multi-volume set called simply *Gray's Botanical Text-Book*. He also spent many years publishing multiple volumes called *Synoptical Flora of North America*.

As his personal and financial fortunes began to improve, he was offered the opportunity to travel and explore in the same vein as his British contemporaries. Indeed, it has been suggested that "(Naturalist) expeditions were the graduate schools of the nineteenth century for many of the best scholars and scientists."

Expeditions Do Not Always Go As Planned

As with most of the other apostles, Gray was also appointed to an expedition. In 1836 he was selected as the botanist to a U.S. exploring expedition to the South Pacific under the command of a Captain Wilkes. Unfortunately, there were many changes and delays, with sailing eventually put off for two years. Tired of the many changes, Gray resigned his position even as the expedition finally started. Having already resigned his post at the Lyceum in order to work for the expedition, Gray was again in need of a job.

The offers for travel were on and off the table mainly because of funding and control issues. Both private societies and the University of Michigan tried to muster the funding. Then in collaboration with the government, the multiple bureaucracies planned trips, withdrew them, and then planned new ones again. In 1838 it was the new State University in Michigan which offered Gray the chair of Botany, a position very valuable no matter the century, but one which Gray never fully undertook. He had negotiated a start date more than a year in the future to enable him to travel to Europe. By the time Gray returned to the United States, funding for the position had evaporated.

Travel in Continental Europe and England

Gray eventually did go abroad as a scholar, but he did not go as part of a crew or as an onboard field naturalist like Darwin and Wallace had done in their prime. In 1838, Gray traveled to Europe in an attempt to regain some of the plant examples he had sent to colleagues years before and also to purchase representative samples of European flora to take back with him to his university.

While in Europe in 1838, Gray had bought books for the brand-new University of Michigan library.[152] He had also visited many herbariums, and had met many fellow botanists. Both of these latter activities proved momentous to his later career and activities. He visited England, Scotland, France, Germany, Switzerland, Italy, and Austria. Among the botanists he met was Joseph Hooker, with whom he began a lifelong correspondence. Through Hooker he also met Charles Darwin, though they only commenced corresponding when Hooker reintroduced them in 1855.

But once they were reintroduced, Darwin wrote to and learned a great deal from Gray, who was a very careful systematic botanist, and therefore capable of explaining to Darwin the details of the distributions and classifications of many plants. Darwin also made a point of asking questions in their correspondence, which pointed up problems in the then-current scientific understanding of the world. Darwin, of course, knew what he was doing. Through this careful, almost Socratic type of questioning, Darwin was slowly preparing the ground so that Gray would accept the thesis of evolution by natural selection.

This is how it came to be that Darwin wrote to Gray the famous letter of July 20, 1857,[153] in which he quietly laid out his bombshell idea. By then, Gray was ready to accept it.

Ironically, while he was in Europe, much of Gray's botanical work consisted of looking at specimens of American plants for his continued work on his books, including three separate volumes each on the flora of North America. Once he returned to the United States, he continued work on this series, being published prior to his career at Harvard University.

In Europe he traveled extensively throughout England, France, Germany, and Switzerland. While this was not "exploration" of the kind like Darwin was doing, it was still important as the samples he sent back helped widen the subject of field biology for generations of pupils, faculty, and scientists within the Americas. In 1839, while in England, he spent time with Hooker who in turn made fast connections between Gray and Robert Owen. Hooker, always the bridge builder, introduced Gray to Darwin and, brokered by Gray with his family legacy at Kew Gardens, the three had lunch together at the Royal Gardens and shared their ideas about geology, botany, and natural history.

The mid-nineteenth century was also a good time for Gray to focus on doing more writing and rounding out his career. While *Elements of Botany* had been published years earlier, Gray got around to writing more after 1840. His popular *Notes on Orchids* was published in the *American Journal of Science*. And in 1842, though connections, he was offered a teaching position at Harvard University as a full professor of Natural History.

Harvard

Back in the United States and while at Harvard, Gray maintained a prodigious output of published materials. In addition to research articles, the *Botanical Text-Book* (which went through many editions), and *Flora*, he published volumes on rare plants in North America and a pair of books called "Botany for Young People." This duo explained how plants grow and how plants interact with their environments. He also published a field guide, a general manual of botany for the northern United States, and the textbook eventually called *Elements of Bot-*

any. As the American West was explored, more and more specimens came into the Botanic Garden at Harvard University, for which Gray was responsible. Reports as to the findings made possible by these new specimens were then published by Gray and his colleagues, including many former pupils.

Gray was in some ways a man dedicated to the thinking of his times. He was a religious scientist but not a biblical literalist. In his early career he had been a defender of Bishop William Paley, who had claimed in his 1802 book *Natural Theology* that just like a watch with all its complexity cannot exist without a conscious designer, in the same way humans and all forms of life must also have a conscious "watchmaker." But Gray had moved on from the "watchmaker" analogy. He accepted and maintained that to understand God as a divine force for good, one only need to look at the complexity of nature.

He would certainly be at odds with fellow scientists like John Draper and T. H. Huxley, who needed no supernatural force to explain evolution or life on earth. But Gray's ideas and goals were to not offend the church while working within science, all to accept the natural world on bifurcated terms—to essentially see a metaphysical world of positive intent operating behind the natural world. This tightrope walk was as difficult then as it is today.

The move to Boston brought stability to Gray's career and finances. It was also a springtime of sorts for his heart. In 1847 Gray fell in love with and married Jane Lathrop Loring, the daughter of a well-connected attorney named Charles Loring. Over the next decade life for the Grays, who did not have any children, was focused on Asa's continued writing, research, publications, and lecture schedule.

Gray and Darwin

In 1857, nearly two years before the publication of *Origin,* Gray and Darwin shared a correspondence. It was not unusual for Darwin to reach out to scholars and throughout his life he built a network of friends and colleagues to share ideas and challenge assumptions. Darwin certainly knew about Gray from his earlier books and writings and from his connection to Hooker. Gray deeply respected Darwin and was proud to be considered a friend to the elder statesman of natural history.

Darwin wrote to Gray in July of 1857:

> [A]s an honest man I must tell you that I have come to the heterodox conclusion that there are no such things as inde-

pendently created species—that species are only strongly
defined varieties. I know that this will make you despise
me.—I do not much underrate the many huge difficulties
on this view, but yet it seems to me to explain too much,
otherwise inexplicable, to be false.[154]

And as a retort, Gray responded to Darwin's confession in a letter
dated August of that same year:

No one can have worked at systematic botany as long as I
have, without having many misgivings about the definite-
ness of species. My notions about varieties are I believe just
what you would have them....

And when we see that every plant man takes in hand deve-
lopes into varieties with readiness, when favorably circum-
stanced, we cannot avoid suspecting they may do the same
thing—.i.e. sport in some way in the wild state also;—and
that there is some law, some power inherent in plants generally prompting them to originate varieties.—which is just what
you want to come to, and I suppose this is your starting point.

Here you begin then with good, tangible facts; and I am
greatly interested to see what is to be made out of them.[155]

Gray was only the third person to whom Darwin had admitted his
"heteredox idea," the two earlier confidants having been Joseph Hooker and Charles Lyell.

When Alfred Russel Wallace wrote to Darwin about his own personal discovery of what we now call evolution by natural selection, Darwin was in the midst of gathering evidence for a massive tome on this
same subject, which had been in the making for twenty years or more.

Wallace's letter threw Darwin, and by extension Hooker and Lyell,
into a panic. Darwin, ever the careful man who did not want to rock
the boat, had been trying to assemble enough evidence, experimental
and observational, to write a definitive, multi-volume treatise on evolution. Knowing how blasphemous his idea was, Darwin wanted to lay
out so much evidence that all arguments about it would be stopped before they could even be started. Because of his desire to keep his ideas a
secret until he could write this comprehensive, unbeatable work, Dar-

win limited who he told about his radical biological thoughts to the aforementioned three.

As a result, when Wallace's letter arrived, Darwin had no *published* proof that he had thought of evolution and natural selection first. Darwin was anguished in two ways simultaneously: on the one hand, he wanted to treat Wallace completely fairly; on the other hand, he did not want to have worked on the subject for over twenty years, only to have someone else publish before he did.

This was where Darwin's letter to Gray came in. Here was a letter, written to an established scientist on the other side of the Atlantic Ocean, in which Darwin had outlined his ideas, and he had written it before Wallace had written to Darwin! What's more, Darwin had followed up his letter with an essay on natural selection, which he had sent to Gray along with another letter on September 5, 1857. Fortunately for Darwin, he had kept a rough copy of that September 1857 letter.

These works, combined with excerpts from the book-length essay that Darwin had written in 1844 that he had allowed Hooker and Lyell to read but had never published, allowed Darwin to honorably claim that he had thought of, and written about, evolution by natural selection before Wallace had.

Thus it was that Darwin and Wallace were able to lay claim to the idea of evolution by natural selection at the same time, at the same meeting of the Linnean Society. Hooker and Lyell managed to present both men's work at once, give both men credit for the idea, present the most important biological idea of all time, and keep peace between all parties.

And Asa Gray's letter from Darwin was a crucial piece of this puzzle.

After two years of correspondence with Darwin and Hooker, Gray would come to accept the science of natural selection, even as it materially conflicted with his religious constructs and beliefs. That's because the scientist within Gray could not deny the long-term work of Darwin, nor could he find evidence to disavow what Darwin had written in *Origin*.

Gray did more than just accept Darwin's ideas. He became Darwin's chief advocate in the United States. Gray secured the American printing rights of *Origin* on behalf of Darwin, and the book was first published in the United States on May 1, 1860.[156]

Gray also worked to publish numerous positive essays on natural selection and made sure other positive reviews and comments from different readers and editors made the daily press as well as the scientific literature. Like his European apostles, he also debated on behalf of Darwin within the United States. Many critics of natural selection in America (as well as Europe) believed that our evolutionary existence could not have happened without a divine spark which served to separate us from beasts and nature and provided our species an ethical conscience. For Gray, evolution had nothing to do with what he called "moral law." We evolved from God's plan and that spark of goodness (according to Gray) lives in each of us while at the same time we evolved from animals and remain so connected to nature.

A Scientific Mystery Solved by Evolution—the Asa Gray Disjunction

In terms of emotional impact, the zoological aspects of evolution by natural selection are gripping and dramatic. Thinking about the descent of humanity from earlier forms of life, the drama of predators versus prey species, or even the evolution of the vertebrate eye, are engaging and even enthralling. Stories about the animal kingdom are emotionally compelling, especially when those animals are sometimes people.

This means that people can often overlook the vast quantity of botanical evidence for evolution. But even here there is drama. The drama comes from mysteries that botanists slowly became aware of in their own work: mysteries of why certain plants grew in some places but not in others; mysteries as to why entire ecosystems changed abruptly, going from one community of plants that tended to grow together to another one with very different members, and within the space of a few miles, even when the physical environments were similar; mysteries as to why habitats that are structurally and climactically similar don't support similar species. And why do communities of co-occurring plants show up in wildly disparate parts of the world? These riddles were the stuff of the quiet dramas in botany in the 1800s. Like detectives in mystery novels, botanists gathered evidence and looked for clues but could not figure out that one missing piece of the puzzle, the piece that would make the whole picture come into focus.

One of these now solved riddles is still famous to biologists to this day, and bears the name of the man who studied it: the Asa Gray Disjunction.

The Asa Gray Disjunction is the name given to the curious fact that there are two communities of co-occurring plants, complete with a whole catalogue of shared species and genera, in regions that are wildly separated, geologically, from one another.[157] What's more, many of the species in these two regions are found nowhere else in the world! The two regions are eastern North America, and East Asia.

By the mid-1800s botanists had come to the astounding conclusion that the plants of eastern North America had more in common with plants in East Asia than they did with the plants of the American West.

Today we know that the shared groups that are peculiar to these two regions include about sixty-five shared genera, a few genera that are closely related, and a few species. Within the shared genera, there are species from eastern North America and East Asia that are extremely similar. Among these are sassafras, pachysandra, dogwood, honey locust, witch hazel, Virginia creeper, ginseng, skunk cabbage, and arborvitae trees. (The latter are confusingly known as the cedars, such as the eastern white cedar, even though they are not in the genus *Cedrus*, like the cedars of Lebanon that Hooker studied, but rather in the genus *Thuja*. *Thujas* are only found in eastern North America and East Asia, while *Cedruses* are found elsewhere.)

The similarities can be striking. Sixty-seven percent of the seed plant genera in Maine also grow on the island of Honshu in Japan. There is also fossil evidence, with extant species in Asia being found in fossil form in North America, and vice versa. What's more, we now know that plants are not the only things that these two regions have in common. There are also fungi, arachnids (spiders and their relatives), millipedes, insects, and freshwater fish that are found only in these two areas.

Gray studied this puzzle and wrote about it but didn't know what to make of it. Then in 1855 Darwin reintroduced himself to Gray by letter, by way of their mutual friendship and correspondence with Hooker. Darwin said he needed help with some plant species distribution problems of his own.

Darwin's need was genuine, and what he learned from Gray over the course of their correspondence strengthened his scientific work. However, there was a human angle here as well. Darwin and Hooker had noticed peculiarities in plant distributions elsewhere that appeared to support evolution by natural selection. For instance, plant species on islands in the Galapagos that were near to one another often had closely similar species, which suggested common descent, just as closely similar species of finches that were on islands in the Galapagos that were near to each other also suggested common descent. These finch species are now famous, and are referred to as "Darwin's finches." Hooker had also noticed that plant communities in the European Alps were similar to those in Arctic, which suggested that they had once shared a habitat, and then become separated.

Hooker already knew about Darwin's theory of evolution from common descent, and also knew about the puzzle that Gray had worked on concerning the plant communities in eastern North America and East Asia. He thought that common descent might be at work in that case as well. He further surmised that if Darwin's theory helped Gray solve his mystery, then Hooker would not only be doing a fellow botanist a favor, it would win Darwin and his theory an important ally.

It worked. By the time that Darwin revealed his thoughts about evolution, Gray and Darwin had come to respect each other as scientists, so the idea was not rejected out of hand. More importantly, Darwin had been plying Gray with evidence for evolution during their correspondence. Gray, an open-minded scientist, was willing to entertain the idea. Then Gray, a careful, show-me-the-data kind of scientist, became increasingly convinced, since the data all supported Darwin's idea.

Crucially, the idea of evolution by natural selection, and therefore the idea of closely related species having common ancestors, made much more sense as an explanation of the mystery of the Asa Gray Disjunction than did the idea of two very similar plant communities in two wildly separated areas of the world being specially created by God. Evolution just made more sense. Like the detective in a mystery novel, Gray saw that evolution was the missing piece of the puzzle that, when in place, made sense of the whole picture.

As with most good mysteries, there were still threads to tie up. For evolution and common descent to work, Gray needed to figure out how these two communities had once been connected and then got separated. Gray figured out that there once must have been a land bridge across the Bering Strait, between East Asia and western North America. This would then provide for the connection between the two plant communities—these plants had once been spread out throughout northern North America and Asia! How did they then become separated?

In one of history's great ironies, Gray applied the new, ground-breaking work of his colleague at Harvard, Louis Agassiz. Agassiz was the first person to suggest that the earth had once had an ice age. He did this by studying extant glaciers and seeing what they did to the land and rocks around them. Then he noted signs of glaciers that had existed but were no longer there in his native Switzerland. These signs included mounds of debris called moraines that glaciers deposit on land in areas that did not have glaciers in Agassiz's time. Evidence also included erratic boulders that had been carried long distances by glaciers and then deposited, scratching and smoothing of rocks, and great valleys dug out by glaciers. In later research he also found this same evidence of glaciation in North America.

So Agassiz had posited that a huge, wintery ice sheet had once gripped the north. This then provided the separation event between Asia and North America that Gray needed. As the world got colder, the plants that had flourished in the north were driven farther and farther south until the American and Asian populations were separated from one another. Then, after the ice age, later cooling and drying (for instance in the American West) forced these plant communities to retreat further, to the cool but not frigid temperate and moist areas where they now thrive. That's how these communities wound up existing to this day in northeastern North America and Japan, Korea, and eastern parts of China.

We now know much more about the periods of glaciation and the warmer periods in between than Gray knew at the time. What I have just described is now described as being that a land bridge between Asia and North America existed during the Miocene geological epoch. This allowed for the spread of species between North America and Asia. During the Quaternary period, cooling and drying took place, and the species in eastern North America and East Asia retreated southward to form the two plant communities that we now see.

Hooker had similarly combined the idea of an ice age with the idea of common descent (as opposed to special creation) to explain the fact that Arctic flora and European alpine flora have many species in common. He figured out that this floral community originally developed in Scandinavia. It was then driven southward by the approaching ice during a glacial period. Then later, after the cold had become less severe, part of this Arctic floral community remained in the high Alps while some of the rest retreated to the Arctic, along with and behind the ice.

So Agassiz's brilliant and ground-breaking idea of an ice age was helpful to his colleagues and to the advancement of science. Too bad the rest of his ideas didn't follow suit, as he was an anti-Darwinist and would never accept evolution as true. This meant he was at permanent war with Gray and the rest of the apostles.

Louis Agassiz—Gray's Colleague and Adversary Who Opposed Evolution

Agassiz was the brilliant man who had thought to study glaciers and who provided evidence that the world had once been subject to an ice age. He did great work in classifying fish species, both fossil and living. He was a terrific popular lecturer. After being born in Switzerland and having an early career in Europe, he was offered a position at Harvard on the strength of a series of wildly popular lectures that he had given in Massachusetts. While at Harvard, he did what ambitious Victorians did: he started a collection. He founded, and did the collecting for, one of the great zoological museums in the United States: Harvard University's Museum of Comparative Zoology. That museum is still a place of active research today. He worked for the support of science in the United States and helped to form the National Academy of Sciences. He was also a regent of the Smithsonian Institution in Washington, D.C.

Agassiz was charming, spirited, a great talker, and at one time America's most famous scientist. He was a scientist who became a celebrity, a Victorian-era equivalent in fame to Albert Einstein in the twentieth century. Unfortunately, Agassiz also believed that a species was "a thought of God." So firmly did he believe that species were ideal creations by God that he was completely unwilling to brook the idea that they ever changed or evolved, and even unwilling to admit that there was much variation within a species. If a species was perfect, why would there be variation? He claimed that "varieties, properly so called, have no existence, at least in the animal kingdom."[158]

This, of course, is not only false but it also conveniently ignoress the entire plant kingdom. He was so firmly convinced that species were fixed that he would not even admit that there was any relationship between Tertiary period seashell fossils and present-day sea creatures. He also believed that different classes, orders, families, genera, and species were either "higher" or "lower" than one another.

Agassiz extended these ideas to human beings. He thought that different races were actually different species that had been created separately, in different places, by God, and that they were not equal. In fact, he classified humans into five different species that were indigenous to specific regions on earth. He felt that the "highest" in development were white Europeans (like himself, of course) and the lowest were black Africans. He was horrified at the idea of racial mixing.

Not only is this disgusting today, it was also seen as disgusting at the time, at least by Asa Gray. In the mid-1850s, slavery was the hottest issue in the United States, and the nation had been on the brink of a civil war about it for some time. Agassiz's beliefs, posing as science, gave an apparent scientific cover for racist arguments in favor of slavery. Interestingly, this not only conflicted with Gray's developing ideas about species descent, it also conflicted with Christian beliefs about the common origins of human beings.

What else was wrong with Agassiz? He cut corners intellectually, if this would make a popular lecture more entertaining. According to

Gray, Agassiz's love of popular lecturing "has greatly injured him" and led him to "tamper with strict veracity for the sake of popular effect." He even went so far as to praise in print, in the *American Journal of Science*, a book which he had not even read! He apparently did so because the book was by one of his protégés and it attacked the work of a friend of Gray's!

The worst part, however, continued to be Agassiz's romantic insistence on an idealized view of nature that did not agree with the facts that Gray kept uncovering, and that agreed with and prized ideals, logic, and even religion over empirically-derived facts, *in science!*

As science was moving in an ever-more empirical direction, Gray realized that it was also becoming more solid and reliable. He wrote to Hooker in 1858:

> [I] sympathize more with & estimate higher the slow in-
> duction that leads step by step to sound conclusions so far
> as they go, than the bolder flights of the genius which so of-
> ten leads the possessor to mount three pairs of steps only
> to jump out of the garret window.[159]

The idea of species as a direct result of God making them, and putting them in exactly the locations where he wanted them, and in exactly the quantities he wanted them, qualified as that sort of a jump out a window. All of these were annoyances and worse to a careful scientist like Gray. But in the center of this was the fact that, for many of the reasons just stated, Agassiz was completely and loudly opposed to evolution by natural selection.

Unfortunately, this could not remain a simple academic spat. Agassiz was famous enough and influential enough that he could crush theories that did not meet with his approval. Getting past Agassiz was crucial if evolution was to be accepted in the United States. But likewise, Agassiz had built so much of his career and scientific reputation on his vision of ideal types—the idea that species are works of God—that Agassiz knew either Darwin went down or he did.

Thus was the stage set for confrontation yet again, this time in the United States. Again a powerful zoologist and collector with money, power, and ambition sought to smash evolution, and a botanist and collector with power, ambition, and at least some money, opposed him. As it was in England, they both worked for the same place. In England, both Owen and Hooker worked ultimately for the British government.

In America, Agassiz and Gray worked for Harvard. In both cases the intellectual and scientific stakes were high, and the zoologist was not above threatening the botanist's livelihood.

In England, Owen tried to have Hooker removed from his post at Kew Gardens. In the United States, Agassiz asked the Massachusetts textile magnate John Avery Lowell to substitute in for him in one of his confrontations with Gray. Substituting a businessman for a scientist in an important scientific meeting, and having him attack Gray on religious and philosophical grounds, seems like a strange and perhaps desperate thing for Agassiz to have done. Strange that is, until you realize that Lowell was a benefactor of Agassiz's, and more significantly, he was the dominant member of the Harvard Corporation—the governing body of Harvard University, which was Gray's employer. Displeasing a member of the corporation could have had a negative impact on Gray's continued career at Harvard, and the university's continued support of the botanic garden. Agassiz's sending in Lowell to do his work for him was a means of intimidating Gray.

Gray's and Agassiz's public confrontations got started when Gray realized that his research on the Asa Gray Disjunction (as it is now known) was a great way to begin to challenge Agassiz. It even used Agassiz's own idea of an ice age! But at the same time, it disagreed with nearly everything that Agassiz had said about natural order and how species are created. The ice age itself was given a more empirical interpretation—it wasn't a catastrophic event that had been used by God to wipe out species and create new ones in their place but rather a natural event that in some cases gradually pushed species around rather than destroying them. It was in general a piece of research and interpretation that favorably contrasted Gray's careful empirical data gathering to Agassiz's flashy vision.

So Gray presented his work on the disjunction at the December 10, 1858, meeting of the Cambridge (Massachusetts) Scientific Society, a club to which both he and Agassiz belonged. The timing here is interesting. It is after the famous meeting of the Linnean Society when Darwin and Wallace's papers about evolution were first presented, but a year before *Origin* was first published in England.

He followed this up by expanding on the paper in a presentation to the meeting of the American Academy of Arts and Sciences the next month. This then became a published paper in the *Memoirs of the American Academy of Arts and Sciences*.

In a footnote to this article, Gray quietly announced that Darwin's theory would explain the "fundamental and most difficult question remaining in natural history" and said that it would have "a prominent part in all future investigations into the distribution and probable origin of species." Thus, although the paper was strictly about the science of Gray's peculiarly dispersed plant communities in North American and East Asia, it quietly and effectively slipped in Darwin's groundbreaking theory in such a way that what was highlighted was its clear and obvious scientific utility rather than its earth-shattering nature. He wrote:

> the limits of occasional variation in species...are wider than is generally supposed, and...derivative forms when segregated may be as constantly reproduced as their originals.[160]

If you read this carefully you can see that Gray is saying that genetic variations can produce new species! So much for Agassiz's fixity of species! Yet this is all done within the quiet context of explaining a botanical puzzle regarding the distribution of plant populations.

In his oral presentation to the Academy, Gray stated explicitly that his paper contradicted the ideas of species fixity and distribution which had been the hallmark of Agassiz's work. Agassiz was present at that meeting. He responded quietly for about half an hour, saying, perhaps reasonably, that he knew zoology better than botany. Unreasonably, he then simultaneously repeated his past positions about species fixity and distribution, without addressing the botanical evidence that had just been presented. The battle was on.

The two men debated throughout the winter and spring of 1859 in meetings of scientific societies and public forums. Agassiz went to Europe for the summer, and in the fall he quietly resumed his duties at Harvard. Perhaps he hoped that the debates were over. Then in November of that year *Origin* was published, and the world changed.

"If Darwin is right, Agassiz is wrong," was the conclusion of an excited group of readers of *Origin* shortly after the book arrived in the United States, around Christmas of 1859. Those readers knew what they were talking about—they included the Harvard zoologist Jeffries Wyman, Agassiz's friend, the poet James Lowell, the Harvard aesthetics professor Charles Eliot Norton, and historian Henry Torrey.

The book was immediately popular, exciting, and intellectual. It was also filled with actual natural evidence. Agassiz recognized the threat and by January of 1860 he was back at it. At the January meeting of the American Academy of Arts and Sciences he asserted again the fixity of

species, and astoundingly claimed that there was no genetic relationship between fossil seashells from the Tertiary period and creatures
that are alive today. More meetings and more assertions followed.

There were multiple reviews of *Origin* itself. Gray wrote a long, careful review in the March, 1860 issue of the *American Journal of Science*.
In it he contrasted Darwin's careful science with Agassiz's ideas, which
were "theistic to excess." This was followed by a three-part article in
the July, August, and September 1860 issues of the *Atlantic*. This is telling, since the *Atlantic* was a popular magazine that appealed to general
readers, and what's more, it was both edited and owned by friends of
Agassiz's. Yet, they allowed not just one positive article on *Origin*, but
three, in successive months. The wind was blowing in a new direction.

Agassiz didn't get around to trying to rebut *Origin* until the July
1860 issue of the *American Journal of Science*. Sadly, he did not really refute Darwin, or even try to. Instead, he asserted that Darwin was
wrong and repeated some of his earlier messages.

His scientific credibility waned but not his energy. Although Agassiz's work as a collector and curator at the museum continued to be valued, his days as a respected theoretician were over. He rarely managed
to publish or lecture in serious scientific settings after that, but he published and lectured anyway, producing twenty-one articles and four
books and giving numerous public lectures between 1861 and 1866,
mostly in popular venues. But his scientific colleagues, and eventually
his students, one by one, started accepting evolution. In time Agassiz
pressed on with his increasingly untenable position, all by himself. The
Victorian equivalent to Dr. Einstein had become the Victorian equivalent to Dr. Oz. His was another unfortunate case of a beautiful vision
being ruined by unattractive facts.

Aristotle's beautiful idea of crystalline spheres in the heavens controlling the movements of all heavenly objects around the earth, ideas that the Catholic Church clung to, were ultimately destroyed by
Galileo's annoying and ultimately shattering facts. In the same way,
Agassiz's elegant vision of Platonic forms, in an ordained order in biological nature that was ruled by God, was also ultimately destroyed by
annoying and unruly biological facts, emanating from Darwin, and
also from Gray.

As with the famous debate at Oxford, sometimes fanciful tales are
told. It is said that at a meeting of the National Academy of Sciences,
which Agassiz had helped to found, a member was elected who was
an ally of Gray's rather than an ally of Agassiz's. Agassiz is said to have

been incensed, and to have challenged Gray to a duel, no less. It is lost to the mists of time whether or not this challenge was actually made. In any case, there was no duel, though it was several years before the two of them would speak to each other again.

So Gray certainly won in the scientific realm, successfully defended Darwin, and paved the way for acceptance of evolution by natural selection by scientists in the United States. Still, the possibility exists that Agassiz was able to do some lasting harm to science, since his popular writing not only gave incorrect but respectable-sounding intellectual ammunition to pro-slavery and pro-racism campaigns, it also gave incorrect but respectable-sounding intellectual ammunition to those in the popular and political arena who refuse to accept evolution.

Agassiz was not completely defeated in the popular imagination—hence our continued troubles with accepting evolution here in the United States. One can hear echoes of Agassiz's "higher" and "lower" ideas every time that a racist argument is made, and echoes of Agassiz's idea of ideal types every time a creationist asks why there are no transitional fossils (when there are many), or why, if we are descended from monkeys, there are still monkeys. A lack of willingness to even consider the idea of common ancestry is basis of those questions. Just as Dr. Oz's nonsense will continue to befuddle some people for many years to come, Dr. Agassiz's hidebound stubbornness and insistence that science conform to his lovely visions will also continue to plague us for many years in the future.

Slavery, the American Civil War, and Evolution

The dark years for Gray and most Americans as this point came during the American Civil War. He taught mainly during this time and began to think about retirement after more than four decades as a field biologist, classifier of plant species, college professor, and lecturer. In 1866 he made an early attempt at retirement in the post-Civil War era. However, with not enough money to secure his later years he continued to work. In 1872 he again tried to leave the university but could not afford to live independently from his faculty salary. Between Gray's first and second failed attempt at retirement he did get a chance to visit his friend Darwin in 1868 at Down House while visiting Europe.

Interestingly, Darwin was also an ardent abolitionist, although he was not, by any measure, a suffragist. He wrote about abolition:

Some few, & I am one, even wish to God, though at the loss
of millions of lives, that the North would proclaim a crusade a crusade against Slavery. In the long run, a million
horrid deaths would be amply repaid in the cause of humanity...Great God how I shd like to see that greatest curse
on Earth Slavery abolished.[161]

Gray wrote, regarding the North and South:

All reason appears to be on one side; all madness, audacity,
and folly on the other.

As the Civil War ground to a close, Gray wrote to Darwin:

You see that slavery is dead, dead.

And Darwin wrote to Gray:

I declare I can hardly yet realise the grand, magnificent fact
that Slavery is at an end in your country.[162]

It is interesting that back in those times evolution was seen as a be-
neficent idea by those who desired equality in human relations. Not
long after the American Civil War, Darwin wrote in *The Descent of Man*:

As man advances in civilization, and small tribes are united
into larger communities, the simplest reason would tell each
individual that he ought to extend his social instincts and
sympathies to all the members of the same nation, though
personally unknown to him. This point being once reached,
there is only an artificial barrier to prevent his sympathies
extending to the men of all nations and races. If, indeed,
such men are separated from him by great differences in appearance or habits, experience unfortunately shows us how
long it is, before we look at them as our fellow-creatures...
This virtue, one of the noblest with which man is endowed,
seems to arise incidentally from our sympathies becoming
more tender and more widely diffused, until they are extended to all sentient beings. As soon as this virtue is honored and practiced by some few men, it spreads through

instruction and example to the young, and eventually becomes incorporated in public opinion.[163]

Evolution by natural selection was enthusiastically adopted by abolitionists on both sides of the Atlantic. You may recall that Huxley's lectures on evolution to English workers were popular as well. A scientific idea that opposed the Great Chain of Being and that said that the enviable position of the aristocracy was not foreordained and all a part of God's plan was likely to be popular with those who were opposed aristocratic or racial privilege.

Objections to Evolution by Evangelical Christians

It is also interesting to realize that many evangelical Christians in those times were not biblical literalists. They did not read Genesis literally, nor did they believe in a young earth. What bothered them back in Darwin's time about evolution by natural selection was the randomness of what we now call mutation. It was the lack of divine purpose in all that genetic variability, the lack of divine purpose in all that breeding and dying, and the lack of divine purpose in the appearance of new organisms. This bothered many other Christians as well.

Publication of *Origin* in the United States

Gray not only gave thorough, scientific, and positive reviews of *Origin*, he also personally saw to it that the book was published in the United States. Representing Darwin, he started by negotiating with the Boston publishing house Ticknor and Fields. However, as is the case today, the piracy of intellectual property was a problem. There were already pirated copies of *Origin* circulating in New York, put out by a firm called Appleton's. Darwin had no knowledge of or control over these pirated copies.

Since Appleton's was clearly willing to publish this controversial book, Gray decided to prevent further piracy by negotiating directly with William Henry Appleton. As a result, Appleton's got the contract for a fully authorized text, complete with an endorsement from Charles Darwin himself. The title included the words "New edition, revised and augmented by the Author." The first print run was 2500 copies, and for this Gray sent Darwin a check for twenty-one guineas.[164]

Later Career and European Travel

The 1870s were good times in Gray's and Darwin's relationship. They both became interested in various types of evolutionary adaptations in plants, from insectivorous plants to plant "behavior" (movement and the like) to fertilization. Gray wrote the popular book *How Plants Behave; How They Move, Climb, Employ Insects to work for them* which was published in 1874 and sold widely. Darwin wrote *The Different Forms of Flowers on Plants of the Same Species* in 1877, and dedicated it to Gray "as a small tribute of respect and affection."

When he published the *Descent of Man* in 1871, Darwin gently, or perhaps wryly, feared that he would "probably receive a few stabs from your polished stiletto of a pen." Gray responded that accepting what Darwin wrote at the end of that book was indeed difficult. However, he said, "Almost thou persuadest me to have been a hairy quadruped, of arboreal habits, furnished with a tail and pointed ears, etc."

Although they never saw eye-to-eye on religion, they were both careful but courageous scientists who worked together with great enjoyment.

Finally, in 1873 Gray did manage to retire, at a time when his colleagues and intellectual foes began to pass away. Gray's longevity made him a standout, a pillar of a time long past, a signpost of sorts representing a gilded age of gentlemanly science.

In 1877, while on a holiday trip to Europe, Gray's presence was clearly lauded. Although this trip was meant purely as a vacation with Gray being by no means anything more than a tourist, his appearance on the continent was met with warm openness and respect from the academic and scientific communities. This had become unbeknownst to Gray and his traveling companions one of his last major excursions to the continent. In 1881 he and his wife visited briefly for a time and again came to Down House for what would be the last meeting between Gray and Darwin, with the latter passing away in 1882.[165]

The Great Western Trip of 1877 with Joseph Hooker

But 1877 was also extraordinary for other trips closer to home. In a more research-focused journey, both Gray and his longtime friend Hooker went along together to explore the American West. Can you imagine the great wealth of knowledge these two men shared about botany and science? How they must have talked about their personal and professional difficulties and successes and how they shaped biology? I suppose their long conversations about the friendships they shared with colleagues, including their work in supporting Darwin, must have been life affirming for both men.

What a wonderful trip that must have been! In 1880 they coauthored their findings for the U.S. Geological Survey in *The Vegetation of the Rocky Mountain Region and a Comparison with that of Other Parts of the World.* Their work was done very much in concert so as to gather data and explore how plants are distributed. By way of researching how plants are distributed, they were able to gather evidence for evolution by natural selection.

Darwin was delighted with the results. He was also impressed with Hooker's reports of Gray's physical vigor on the trip to the Rockies. Darwin had previously written to Gray in January 1878:

> I have just spent a delightful 2 hours at Kew, & heard prodigies of your strength & activity. That you run up a mountain like a cat![166]

Why Botany?

Why were botanists relatively willing to grapple with and eventually accept evolution while zoologists were more reluctant to do so? Why was botanical evidence so important to Darwin?

Perhaps it was easier for botanists to accept evolution because we humans have a natural tendency to identify with animals. We may love them or be disgusted by them but one way or another we have emotional reactions to them. By contrast, it is easier to be dispassionate about plants.

It is also easier to see and accept that a common ancestor can produce a wide variety of plants, even within a species. As Gray pointed out:

> Who would have thought that the cabbage, cauliflower, broccoli, kale, and kohlrabi are derivatives of one species, and rape or colza, turnip, and probably rutabaga, of another species? And who that is convinced of this can long doubtingly hold the original distinctness of turnips from cabbages as an article of faith? On scientific grounds may not a primordial cabbage or rape be assumed as the ancestor of all the cabbage races, on much the same ground that we assume a common ancestry for the diversified human races?[167]

The point here is that we know and fully realize that vegetables as visually different as broccoli, cauliflower, cabbage, and kohlrabi (as well as red cabbage, bok choy, and Brussels sprouts!) are all in the same *species* of plant, yet they look remarkably different. Although we can also see great differences in breeds of dogs as well, we generally find it less threatening when plant species develop multiple varieties, which may in time become species themselves. Blurry edges to the species idea in plants bother us less than contemplating this in fellow members of the animal kingdom. Likewise, contemplating species hybridizing when they are oak trees bothers us less than when they are animals.

It is also easier to gather data on plants. They don't hide, they don't run away, and they hold still while you count and measure them. There are also, frankly, more of them. Any given acre of soil will support only a few vertebrates, a greater number of insects and worms, and thousands of plants. The plants have to significantly outnumber everything else, since they are ultimately the food source that supports everything else. A few animal predators eat a greater number of animal herbivores, who depend on much larger numbers of individual plants to eat in order to survive. Thus a botanist can have tens of thousands of data points for every one that can be gathered by someone studying animals. Recall that Paul de Chaillu was a celebrity in London because he said he had had *seen* gorillas in the wild, and was able to bring back a few dead specimens.

By contrast, Gray and Hooker were able to see and work with tens of thousands of botanical specimens, in the field and in herbariums.

It is easier for botanists to gather plants, measure them, and compare them to one another, in large quantities. What's more, although it is true that animal skins and skeletons can be preserved, it is more difficult and expensive than preserving plant specimens. Plants can be pressed and dried and stored in herbaria relatively easily, so botanists can revisit their own and each other's specimens at later dates, multiple times.

Therefore, for botanists like Gray and Hooker, the sheer weight of their data would tend to drive them to support evolution by natural selection. In fact, in the history of science, sheer weight of data in general tends to elucidate patterns and create new knowledge. When a new means of measuring something more precisely, or a new way of observing something more accurately, is developed and used, new knowledge, new patterns, new answers, and new questions almost inevitably fall out as well.

Later Writing and Legacy

In the decade between these last major trips, Gray's retirement was spent working quietly and privately. He kept a busy schedule and became what many would consider the American version of an English country gentleman. He spent his time seeing both family and friends, meeting with colleagues, and remaining relevant in retirement as a storehouse of knowledge regarding botany. It was a fulfilling decade in his life as his work pace slowed although his interests and commitment to science, and his faith, never wavered.

In 1876 Gray published a book of essays titled *Darwiniana*. In this book he attempted to reconcile his own Christian faith, which he retained his entire life, with his acceptance of evolution. His conclusion was that both faith and acceptance of evolution could be held by the same person, at the same time, in a world view that is now called theistic evolution. Gray's version held that mutations did not always arise by chance but rather were at least sometimes generated by God, though after that, natural selection was able to take its course.

A more modern version of theistic evolution simply holds that God existed at the very beginning of everything, gave the universe a push to get it started, and then got out of the way. As a sort of ultimate cause that happened even before the Big Bang, it falls well outside of any investigation by science. It is not science but it doesn't get in the way of science either.

In any case, Gray insisted that accepting evolution did not require either belief or disbelief. He insisted that science is neutral.

Gray is sometimes referred to as "Darwin's Dove"[168] (as opposed to Huxley being "Darwin's Bulldog") because he promoted peace and reconciliation both around good science and between science and religion.

However, it is notable that Gray required religion to conform to science. Religion had to concede to facts established by science instead of truth being obscured in order to uphold a religious vision.[169] This was a long way from the how science had been hamstrung in the early 1800s and before, when "science" had largely been practiced (in England at least) by gentlemen members of the clergy who had made sure that scientific "facts," and the conclusions drawn from them, always conformed to church doctrine. By the time that Gray wrote his conciliatory pieces in *Darwiniana*, that ship had sailed.

In the end, Gray accepted evolution because he was a careful scientist. Being a careful scientist, Darwin's amassing of multitudes of data impressed him. More to the point, Darwin's theory did for him what a good theory does. That is, it answered a lot of questions that were otherwise unanswered, or even unanswerable, while also generating many hypotheses that were testable.

Darwin's data were both observational and experimental—that is, Darwin did the hard work himself of testing some of the hypotheses that were generated by his own theory. As those hypotheses were tested, the evidence for evolution became firmer and firmer. The two men were united by a fascination with nature, a willingness to do the hard work of uncovering facts in nature, a willingness to look facts in the eye, and by agile brains able to see larger pictures emerging from many small facts. Is it any wonder that they worked well together, and in so doing, became lifelong friends?

In late November 1887, Gray became seriously ill. He suffered from paralysis, which robbed him of his ability to speak or move. Illness had robbed this great man of his faculty for communication and the thing he perhaps most prized, the need to be amongst people and a part of life. After nearly a three-month struggle, Asa Gray passed away on January 30, 1888, leaving behind a legacy in the natural sciences and theology and a kind and true advocacy for Charles Darwin and the understanding of natural selection.

While Gray did not have any children, his long work history in the fields of science and botany serves as a legacy that lives on to this day. William Hooker, a son of apostle Joseph Hooker, who was also an esteemed botanist, named a genus of plants "Grayia" in Gray's honor.

Stony Brook University established the Gray Arboretum and two mountain peaks, one in New York State and one in Colorado, each are named after Gray. In 2011 the United States Postal Service honored Gray with his own postage stamp. Since the United States does not put scientists on its currency, like Darwin on the U.K. ten pound note, this official national commemoration alternatively and effectively does honor to Gray's contributions to science in general as well as to botany and the acceptance of natural selection and evolution in particular.

Gray is also commemorated at Harvard, the institution that he served for so long. The Gray Herbarium is part of the university's greater herbaria and is composed of specimens that Gray obtained, sometimes getting them himself and sometimes obtaining them through exchanges with other botanists like Hooker. The Harvard Botanic Garden, though not named after him, flourished and rose to international prominence under Gray's directorship. The Asa Gray Professor of Systematic Botany is a position at Harvard named in his honor.

The Asa Gray Garden, part of the Mount Auburn cemetery in Cambridge, Massachusetts, is also named in his honor.

CHAPTER 10

JOHN WILLIAM DRAPER

A true theory is like a window of crystal glass, through which we can see all objects in their proper positions and colors and relations, no matter whether they are such as are near or those that are at a distance; no matter whether they are directly before us or enter only obliquely into the field of view. A fictitious theory is like a Venetian blind, which has to be set in a certain position with respect to the observer and only shows him objects for which it has been adjusted, and those in an unsatisfactory manner; but if he moves to one side or to  *the other, or endeavors to see objects which are not directly in his way, his view is intercepted, or, perhaps, unless he makes a new adjustment, the light is shut out altogether.*

John W. Draper

Briefly:

John William Draper immigrated to the United States from his native England. He was a chemist and would eventually become the first president of the American Chemical Society. Draper was the first person to ever photograph the moon. In 1841 he helped found the NYU Medical School. Darwin was one of Draper's personal heroes. In 1860 Draper traveled to the United Kingdom to deliver a paper to the British Association for the Advancement of Science that supported evolution. His 1863 book on the intellectual development of Europe placed organized religion and the church as the main barrier to scientific knowledge and discussed the essential philosophical and historical conflict between science and religion.

Draper's Early Life and Career

Darwin and Draper were contemporaries, with Darwin two years older. But the two would pass away only about three months apart in 1882. Draper was born on May 5, 1811, in Lancashire, which is far from London on the northwest coast of England. Draper was not born into the aristocracy and, unlike Darwin, was not connected to great family wealth. However, like many of the apostles, Draper grew up within a religious household. Draper's father served as clergy and because of his need to administer to many flocks in the sparse British countryside, the family moved frequently across the nation.

As a consequence of the constant disruption in family life, young John did not attend any formal school until he was about twelve years old, the family choosing to home school the young boy instead of having him begin and then need to be pulled from the various schools that a clergyman's son would attend at the time. Draper's father would pass away before the son would complete his studies at university. But it is clear that Draper's father approved and supported his young son's interest in science, even buying him a telescope and urging him to seek out a career in the natural sciences, primarily as a physician.

Although Draper's path to a university education did not include much formal school training until he was almost ready for higher studies, Draper, at eighteen, was accepted into the University College of London. At the university he chose to study the sciences, first as a pre-med student and then with a major in chemistry. Because Draper's university was considered less prestigious than Oxford and Cambridge, Draper's diploma at graduation was a "Certificate in Chemistry" rather than a formal degree. The only institutions at the time that could legally offer degree titles were Oxford and Cambridge.[170]

At university Draper came under the tutorage of famed British organic chemist Edward Turner. Turner's lectures and time spent with Draper changed the young man's life. In fact, it was Turner who focused Draper's studies on the chemical reactions related to light. At the time neither man could know that this area would lead the young man into a life-long interest in photography (a technology not yet invented), with Draper being one of the earliest to use the photo-chemicals to document and photograph the earliest portraits as well as phases of the moon.

There are times in one's life that can only be explained as tumultuous, and for Draper that must be how the years 1831 and 1832 can best be described. During this time he collaborated on several scientific papers, got married to Antonia Cetana de Paiva, a Spanish court physician's daughter, and immigrated to the United States. He left England thinking he had a job waiting for him. He and his wife were joined by his mother and three sisters for the journey. It must have been quite a lot to handle for the newly-married twenty-one year-old.[171]

He moved to the United States at least in part because a Wesleyan denominational college in Virginia had indicated that it would offer him a professorship. But as with Asa Gray and the University of Michigan, it turned out that the job offer was not as solid as one would wish. After having moved all the way across the Atlantic with his entire family, the small college told him that he had been so delayed in arriving that they had given his job to someone else.

Nonetheless, even without a job, Draper set up a laboratory. His first solo paper was published soon thereafter in 1834 on the subject of capillarity.

Capillarity is the tendency of many liquids to be drawn up through tubes of very narrow diameter, even in opposition to gravity. This can occur in glass tubes of very narrow diameter, and it is also what happens when fluid is drawn up through the spaces in the fibers of a towel, other cloth, or a wick. Capillarity continues to be an important subject, and Albert Einstein's first-ever published scientific paper, in 1901, was on capillarity as well. Draper continued studies related to capillarity throughout his career, studying for instance how capillarity affects the movement of sap in plants and blood in animals.

Medical Education and Early Science

Draper published four other papers before starting at the University of Pennsylvania. There he studied medicine, as did Darwin and almost all the other apostles. As immigrants have done throughout time, the members of the family depended on one another. The money for Draper's medical education came from his sister Dorothy Catherine's earnings as an artist and art teacher.

While in medical school, Draper continued to be interested in chemistry, and his medical school thesis (titled *Glandular Action*) looked at how gases passed through membranes without being held back in any way, even though the membranes had no *visible* pores.[172]

This illustrates the extent to which the science of chemistry was in its infancy at the time. Although John Dalton had already proposed atomic theory, people did not fully comprehend how small atoms and molecules are. So the fact that molecules of gases could pass through a membrane unimpeded was a new and unusual notion.

Likewise, other concepts that are central to modern chemistry had not yet been developed. Although Dalton had by that time published his ideas on how atoms combine into molecules in certain ratios, Draper started his work in a time before Mendeleev had developed the periodic table of elements. So although there was plenty of interesting work in chemistry to be done, and new facts were emerging at a tremendous rate, organizing that knowledge into a coherent framework was still difficult.

Academic Career

In 1836 Draper got an appointment at Hampden-Sydney College in Virginia. Then in 1839 he moved to the University of the City of New York. Although his original appointment was as a professor of chemistry and natural philosophy in the undergraduate college, the following year he helped to organize the medical department there. This was the beginning of New York University's medical college, which, as NYU Medical School, continues to be one of the top medical schools in the nation. Draper was one of the founders and a professor of chemistry there. He became president of the medical college in 1850.[173]

By 1840, with funds secured, Draper helped found NYU's medical school and served as anchoring professor for a decade. Draper would be deeply committed to NYU for the rest of his life. In 1842 Draper coauthored an important modern textbook, *Elements of Chemistry,* while teaching at NYU. Starting in 1850 he served as the medical school's president for almost twenty-five years. He left the post in 1873 but stayed on as a professor of chemistry, vacating his faculty position in 1881 only about a year before his death.

Not long after his appointment to the medical college, Draper worked with Samuel F. B. Morse, showing that electric current will continue to flow even if the wires are very long. This enabled the development of the telegraph as an instrument of long-distance communication.

In addition to his research work, Draper published several books on scientific subjects throughout his career, including textbooks on human physiology, chemistry, and natural philosophy. (Natural philosophy is what we now call science.) His 1856 physiology textbook is notable for adhering to a strictly scientific view of the human body, eschewing all ideas of a vital force.

The Chemistry of Light

The area of chemistry in which Draper made his greatest contribution was the subject called "Radiant Energy." Today, we would call it the chemistry of light.

One side note of interest in all this is that light was recognized as a form of energy. Today we talk easily of food containing energy, fossil fuels containing energy, electricity being a form of energy, heat being a form of energy, and water running downhill as a form of energy. Having all of these things lumped together in the human mind as forms of energy is a remarkable human achievement and one which is fairly young. It was only in 1844 that William Robert Grove suggested this, and it was only in 1847 that Hermann von Helmholtz came to a similar conclusion and published it as *Über die Erhaltung der Kraft* (On the Conservation of Force), which we now call the conservation of energy.

Light interacts with the physical world in numerous different ways, and Draper studied many of them. He studied how sunlight affects plants, the process which we now call photosynthesis. He studied how light can be produced through chemical action, and significantly, he studied how light can be produced by heat. Many substances, for instance most metals, will glow when they become hot enough. Most people know this, having seen pictures of red-hot glowing iron being worked by a blacksmith, or having seen pictures of glowing red-hot lava, which, when it cools, becomes non-glowing rock. However, figuring out the temperatures at which substances will glow, and how much more light they will emit as the temperature is raised even higher, was one of Draper's avenues of study.

Astonishingly, he described in an 1847 paper a process by which he ran electricity through a thin thread of platinum, gradually raising the temperature by increasing the amount of electricity that flowed through it. This in turn increased the amount of light being emitted by this tiny piece of platinum. He was mostly interested in the spectrums of the light it produced when it glowed at various different temperatures, and how these compared with the spectrum of sunlight.[174] However, he noted that a strip of platinum could reliably be raised to a specific temperature by running a constant current of electricity through it, and that this would yield "a light suitable for most purposes." In short, Draper had figured out the basis for the electric light bulb in 1847, more than thirty years before Thomas Edison patented his model.

Draper also studied how various chemical substances change when they are hit by light. These changes to molecules that take place when light hits them are the basis for vision, and are also the basis for photography. Draper started by studying how metallic salts changed color when light hit them. Then Louis-Jacques-Mandè Daguerre announced the invention of photography by way of the Daguerreotype method.

Draper made a study of it and greatly improved the process. He set up a studio in Washington Square and in 1839 or 1840 took what was probably the first human photographic portrait made in the United States. It was of his sister Dorothy Catherine.

Dorothy Catherine Draper

Throughout his life, Dorothy Catherine continued to be interested in science and gave him considerable help with his research. It is not a coincidence that hers is the first-ever daguerreotype photographed image of a female face. She helped Draper with his research in photochemistry!

In addition she was determined that her brother would be able to fulfill his scientific ambitions. As stated earlier, it was her money, earned by giving lessons in drawing and painting, that allowed her brother to attend medical school at the University of Pennsylvania.

It is reasonable to wonder if Dorothy Catherine might, in a different era, have been a famous scientist in her own right. But this was in the years before it was widely thought possible for women to be independent scientists. Even Marie Curie was not born until 1867, and Dorothy Catherine was an adult in the 1840s.

Other Contributions to Chemistry

Some other specific achievements of Draper's deserve mention.

The Draper Point

This is another achievement related to light, and related to his work that foresaw the development of the electric light bulb. Basically, the Draper Point is the temperature at which a heated solid object will begin to glow, emitting light that can be seen. The glow will be a dim red, and the temperature for all solids will be about 798 degrees Kelvin.

The Grotthuss–Draper Law

This law, developed independently by both Christian J. D. T. von Grotthuss and Draper, states simply that for light to have a chemical effect on a material, that light must be absorbed by that material. So, it doesn't matter what type of light you shine at an object. If that object cannot absorb those wavelengths of light, then the light will not have any effect on it. That is, if the light is not absorbed it simply reflects away with no effect.

But if it is absorbed by a particular material, then the molecules in that material can be changed by the light, causing photographic effects, for instance. When photography was based on changes in chemicals based on light hitting those chemicals, which chemicals absorbed what wavelengths of light was something of immediate importance.

Tithonometer

Related to the Grotthuss–Draper law was Draper's development of an instrument called a tithonometer. This device made it possible for scientists to measure the amount of light that was absorbed in any particular chemical reaction.

The Broad Scope of Draper's Acheivements

Draper was indeed a true Renaissance man. His teaching methods were strong; his ability to build and fund major academic programs made him an asset for the institutions where he worked. His interest in photography would also make him a world-famous artist. Because of his contributions in the creation and preservation of the visual image, something we take for granted now because the accessibility of digital cameras within our phones and mobile devices, we have the first nascent photography created by humans.

In 1839 Draper was as deeply committed to his medical training as he was to early photography. His understanding of chemicals and how they produce a light image dated back to his early years while studying in England. With the knowledge of how chemicals react and the drive to be one of the first persons ever to make a photographic portrait, Draper was a man seeking to advance both science and art (while also making history) as one of the world's first photo portraitists.

As early as 1837 he was experimenting with taking pictures through the process of capturing images onto chemically treated surfaces. But these early attempts did not make or leave any permanent markings, so sadly none survive.

If Draper holds the title of "second person" ever to capture a photographic portrait of a living individual, he also holds a place in science history as the first person to ever photograph the full moon. His work has directly led to centuries of amateur and professionals interested in cosmological photography.[175] We can also link his founding work to the scientific photographic study of the cosmos—first made by humans and now by machines blazing outside of our solar system. With this he helped to initiate the field of astrophotography, which became a crucial tool in astronomical research, and continues to be one to this day.

Draper was so impressed with his photography of the moon that he sent one of his photographic plates to his colleague, the great British scientist, astronomer, and mathematician Sir John Herschel. This image caused a great stir and had come to be revered in the United Kingdom in scientific circles. In the United States he shared the first images of the moon at New York's Lyceum, which was a renowned center for the study and discussion of intellectual, scientific, and philosophical trends. Here too his photographic plates were lauded.

With all this work going on, Draper was a well-regarded man of science with an international set of admirers and connections. He had well maintained a cutting-edge reputation within several fields of inquiry while earning a respected career as a chemist, scientist, professor, and physician.

Draper and Evolution

Like all great men and women of science, John Draper had many influences, just as he influenced generations of scientists himself. Chiefly, Draper was an admitted proponent of Darwin and his ideas, reading Darwin's early works with enthusiasm. Draper was an antagonist against religious doctrine and the organized church. Although prior to the publication of *Origin,* Darwin's and Draper's communication had been limited, Draper was an early defender of Darwin upon the publication of *Origin* and well afterward. In fact, Draper's defense of Darwin rested on two continents, at home in the United States and abroad in the scientific circles in which he also ran while in England.

Draper's defense of Darwin went well into his limited retirement years. His two later books related to scientific discovery, *A History of the Intellectual Development of Europe* (1862) and the *Conflict Between Religion and Science (1874),* both published after *Origin,* have been considered strong philosophical guides with regard to scientific naturalism. Embedded in both books is the idea that the natural world can and should be thought of as the place where humans can best understand the nature of the universe and our human place on earth and in the cosmos.

Although sometimes left unremembered, Draper was on hand for the Huxley-Wilberforce debate of 1860. He delivered a speech on the nature of intellectual evolution within European society. His speech was an unwelcomed undercard event and was not well received at the time of its presentation. Unbeknownst to those in the audience, the speech served as an early version of the book on the same subject to come out just two years later. And while the title of the speech seemed uneventful, it was clear that Draper was a ready and willing soldier in Darwin's intellectual army.[176]

The Famous 1860 Oxford Meeting of the British Association

Did no one tell Draper that he was walking into a hornets' nest? Draper knew and respected Darwin's work and was deeply supportive of natural selection.

Draper knew how religion had hobbled scientific investigation in the past and now threatened to do so in the present. He would have none of it. But did he know that religion was going to intrude on a meeting of scientists in Oxford?

The British scientists at that Saturday meeting all knew each other, or at least knew of each other.[177] Many corresponded or had been instructors and students or had mentors in common. That included Richard Owen and his clerical mouthpiece, Sam Wilberforce. It was a small society.

All the Brits knew that Wilberforce was out to "smash Darwin." All the Brits knew that members of the clergy would be there in force, ready to be raucous. All the Brits knew, whether they liked it or not, that this meeting was going to be the intellectual equivalent of a World Wrestling Federation tag team event, except that no one knew in advance who was going to win.

The seating! The crowds! So many people showed up so long before the official start of the meeting that the organizers of the event had to switch to a larger room. Even still this larger room was suffocatingly crowded, with over seven hundred men and women in attendance.

The British-born and raised Draper had been laboring for years in what he felt was the relative obscurity of the United States, away from European centers of learning. And now he could look forward to high-level intellectual discussion at an ancient British university, with himself at the center of it! He was presenting at the annual meeting of the famous British Association for the Advancement of Science. He was the first speaker on the best day, and with the largest audience! Did he think that all those people had shown up to hear *him*?

What did he think, when he was done speaking, and the moderator of the event had to insist that only people with valid scientific arguments would be allowed to speak—and then nobody listened, and things got really weird? First, someone rushed up to the front, grabbed some chalk, and wrote two Xs on the blackboard saying, "Let this Point *A* be man, and that Point *B* be the monkey," and then all the undergraduates in the room started shouting "Monkey! Monkey!" This kept going until the man at the blackboard was shouted down.

So much for high-level intellectual discussion at an ancient British university!

Then the moderator again demanded that only scientific arguments be made—and who should speak next but a *bishop*, of all people? Why would a bishop speak at a scientific meeting, in his capacity as a bishop? It was known that Wilberforce had never done any biological research.

Had anybody warned Draper that this was going to be a circus, and a circus about whether or not to even *accept* evolution? Draper had written well in advance the paper he was going to present. In fact, it was the subject of a book that he would publish in 1863. Originally, the paper had been titled "On the Possibility of Determining the Law of the Intellectual Development of Europe." When he found out that the subject of the meeting, however it had been heralded originally, was actually going to be about evolution, he changed the topic slightly to "The Intellectual Development of Europe (considered with reference to the views of Mr. Darwin and others) that the Progression of Organisms is Determined by Law."

He was invited to the June 1860 meeting long before it took place, on the twenty-first of March of that year, by Henry J. S. Smith, who was secretary to the Local Committee of the British Association. What a fine opportunity it must have seemed to Draper to expound on his visionary ideas.

Draper accepted evolution, of course. That much must have seemed obvious to him. The great question in his mind was that if biological species' origins could be scientifically analyzed and predictions made using evolutionary theory, then could human events and history also be examined scientifically? Being a historian, Draper wanted to analyze human affairs. Being a scientist, he wondered if there were natural laws that governed human affairs. As a scientist, such a question must have seemed enticing, and potential answers could have heralded a newer, better future! And as a chemist, he may not have understood how the science of biology is different from history. It was an age of rising science, and Draper was not the only person who thought that human politics and history might one day be understood and managed through scientific means. The eminent Victorians Karl Marx and Friedrich Engels thought the same thing. Draper made no suggestions that day for how future societies should be structured, but he, along with others in that era, thought that the forces in history could be studied just like the forces in physics or chemistry.

The man was a highly accomplished polymath who performed wonders in chemistry, particularly in studying how light interacted with anything—chemicals, plants, photographic paper, you name it. He taught at a medical school, wrote a textbook on physiology, and wrote lengthy, respected treatises on history. But he was not a zoologist or a botanist and did not cite biological evidence. He may not have known what biological evidence for evolution really looks like.

As for religion, he had no religious objections to evolution because he considered science to be entirely superior to religion. In fact Draper, of all the apostles, was the most open about being well and truly opposed to much of organized religion. Even Huxley simply wanted religion out of science, and considered himself to be an agnostic. Darwin slowly became an atheist, but not because he wanted to be—he slowly relinquished his faith only because the evidence led him in that direction. But Draper was another matter. He firmly stated that religion actively suppressed science. His criticism of religion was directed at the Roman Catholic Church in particular, and he and his family appear to have remained nominal Protestants. But in 1874 he wrote a book titled the *History of the Conflict between Religion and Science*. It is notable that he did not call the book the *Conflict Between Roman Catholicism and Science*. Although his views of both Protestantism and Islam were kinder than his view of Roman Catholicism, he nonetheless directed the opprobrium of his title to include all religion, not just one in particular.

In this book he popularized the "Conflict Thesis," which asserts that there is a perpetual antagonism between religion and science because religion is constantly defending its antiquated tenets against the new knowledge that is constantly being brought to light by scientific inquiry. He claimed that religion has always opposed the progress of science and has always sought to thwart this progress by every means available

So imagine how bizarre it must have seemed to Draper, when a bishop was allowed and even encouraged to speak at a scientific meeting, and people took his words seriously!

Responses to Draper's Talk at the British Association

Did Draper think that his talk had bombed?

On the one hand, people said nice things to him to his face at the time. After his talk Robert Chambers came up and congratulated him on the "strength" of his presentation. Draper reported that "I was listened to with the profoundest attention" and that "I have accomplished very thoroughly all that I came here for."[178] Afterwards, the editor of the *Athenaeum* asked him for an abstract of the talk so that it could be printed in that journal, and the editor of the *Spectator* offered to print a review of the speech. In a letter to his family back home he wrote that "I cannot now tell you all that has taken place but I may truly say that I [have] never undertaken anything before which so thoroughly succeeded." On the other hand, behind Draper's back, the Brits were not so kind. English historian John Richard Green whined about his "hour and [a] half of nasal Yankeeism."

Joseph Hooker himself complained about

> A paper of a yankee donkey called Draper on "civilization according to the Darwinian hypothesis" or some such title was being read, & it did not mend my temper; for of all the flatulent stuff and all the self sufficient stuffers—these were the greatest, it was all a pie of Herb^t Spenser & Buckle without the seasoning of either—[179]

Spencer and Buckle were in a way popularizers of Darwin's ideas. However, their work often veered away from strict scientific understanding, with Spencer grafting natural selection onto his pre-framed ideas about the evolution of culture, even coining the term, "survival of the fittest."[180]

Years later, in 1898, the writer Isabella Sidgwick recounted the episode in a piece called "A Grandmother's Tales" in *MacMillan's* magazine. In it she made fun of Draper's presentation, saying, "I can still hear the American accents of Dr. Draper's opening address, when he asked, 'Air we a fortituous [sic] concourse of atoms?'"[181]

According to Leonard Huxley, Thomas Henry Huxley's son, Draper's contribution amounted to this: "Dr. Draper droned out his paper, turning first to the right hand and then to the left, of course bringing in a reference to the *Origin of Species* which set the ball rolling."[182]

Did these types of snarky comments completely evade Draper's notice? Even if the debate only became famous in hindsight, Draper's having appeared in Oxford and spoken about evolution before a crowded room at a meeting of the august British Association for the Advancement of Science most likely would have been something that he would mention. Yet even though Draper later wrote an entire book on the conflict between science and religion, in it he never mentioned this interlude, or his role in it.

He also did not mention it when he published his book on the topic of his talk that day—the book titled *A History of the Intellectual Development of Europe,* nor did he bring it up even when he lectured on evolution in the United States! None of his contemporary biographers mentioned it, nor did those who wrote his obituaries at the time. Again, why the later reticence over something that he crowed about at the time, when he said that "I [have] never undertaken anything before which so thoroughly succeeded"? Perhaps word eventually got back to him that the Brits had not thought his work to be so brilliant after all.[183]

It is odd that Draper did not mention his star turn on the stage at Oxford. Despite this, and despite the fact that he was not a member of Darwin's inner circle of friends and advocates in England, he became a key apostle in the United States, along with Asa Gray, for accepting evolution and for spreading the word about Darwin's new theory. He also had plenty of other impressive accomplishments to keep him busy.

First President of American Chemical Society

In 1874 a number of American chemists made a pilgrimage to the grave of Joseph Priestley in Northumberland, Pennsylvania. They did this because exactly one hundred years earlier he had discovered the element oxygen. In addition, Priestley generally helped to lay the foundation for the modern field of chemistry. The chemists celebrated the first century of their field's existence, but noted that although there was a general scientific society in the United States, the American Association for the Advancement of Science, there was no association specifically for chemists. The field of chemistry was burgeoning at the time but each individual chemist worked in isolation. Following the Northumberland get together, chemists in New York decided that chemists needed their own scientific organization and they set about founding the American Chemical Society, which came into being in 1876.

They elected John William Draper as their first president, since at sixty-five he had a long list of impressive achievements and had both national and international standing. Although it was agreed by all parties beforehand that he would not actually participate in running the organization, his stature helped to elevate the fledgling organization, and gave extra flourish to their inaugural events, which were timed to coincide with the United States' centennial celebrations. He gave their inaugural address in 1876. The original band of thirty-five chemists who founded the organization has since grown to over 150,000 members, and the ACS publishes over fifty-three journals.

The Freethought Movement

Darwin's impact on the freethought movement has been deep and it remains both long and wide. But just as Huxley, ever the advocate and activist would work for social change, Draper too, had his admirers and accolades as well.

> Thousands of ministers are anxious to give their honest thoughts. The hands of wives and babes now stop their mouths. They must have bread, and so the husbands and fathers are forced to preach a doctrine that they hold in scorn. For the sake of shelter, food, and clothes, they are obliged to defend the childish miracles of the past, and denounce the sublime discoveries of to-day. They are compelled to attack all modern thought, to point out the dangers of science, the wickedness of investigation and the corrupting influence of logic. It is for them to show that virtue rests upon ignorance and faith, while vice impudently feeds and fattens upon fact and demonstration. It is a part of their business to malign and vilify the Voltaires, Humes, Paines, Humboldts, Tyndals, Hæckels, *Darwins,* Spencers, and *Drapers,* and to bow with uncovered heads before the murderers, adulterers, and persecutors of the world. They are, for the most part, engaged in poisoning the minds of the young, prejudicing children against science, teaching the astronomy and geology of the bible, and inducing all to desert the sublime standard of reason.[184] [Italics added]
>
> These orthodox ministers do not add to the sum of knowledge. They produce nothing. They live upon alms. They hate laughter and joy. They officiate at weddings, sprinkle water upon babes, and utter meaningless words and barren promises above the dead."

—Robert G. Ingersoll, *Mistakes of Moses—The Destroyer of Weeds, Thistles And Thorns, Is A Benefactor Whether he Soweth Grain Or Not* (1880)

The American freethought movement adored Draper, and it is entirely likely that the freethought movement's ideas strongly affected him.

Members of the freethought movement asserted that God did not exist. Some Freethinkers simply considered God's existence to be highly unlikely, others considered that nonexistence to be a certainty. In any case, they believed that all gods, scriptures, and religions were developed by humans, with no divine interaction. They also believed that government and religion should be entirely separate, and that holy books were not good sources for moral values. Rather, they held that good morals could be derived from human experience.

They also held that the only world that matters is the real one that we live in, not any heaven, paradise, or promised afterlife. Science was understood to be far and away the best method for investigating reality, and Freethinkers were decidedly pro-science, as science was a fairly reliable way of obtaining knowledge, far superior to any religious "revelation."

Although skeptics of religion have existed for as long as religion itself, this movement traced some of its history back to the founders of the United States. Thomas Paine, Ethan Allen, Thomas Jefferson, and John Adams all believed in Deism, [185] and while Deism did allow for a God to exist, it also maintained that this God set the universe in motion in an orderly way and was not involved in the world after that.

In fact, a rational, skeptical view of religion was central not only to what many of America's founders personally thought to be true, they also thought that skepticism about religion was crucial to a well-run, civilized society. For instance in 1813, when Britain finally repealed a statute that had made denying the Holy Trinity a crime, John Adams wrote the following to Thomas Jefferson:

> We can never be so certain of any Prophecy, or the fulfillment of any Prophecy; or any miracle, or the design of any miracle as We are, from the revelation of nature i.e. natures God that two and two are equal to four. Miracles or Prophecies might frighten Us out of our Witts; might scare us to death; might induce Us to lie, to say that We believe that 2 and 2 make 5. But We should not believe it. We should know the contrary.[186]

In other words, Adams, despite his reputation for being a Puritan, basically said that you should never believe anything that a religion says if it is contrary to what people can observe in the natural world.

Deism fell out of favor during the course of the "Second Great Awakening," but freethinkers revived their movement in the 1820s.

This time God was rejected entirely by most adherents. Newspapers were founded and public lecture tours were successful. Much of this was focused around New York and the lands served by the Hudson River and the Erie Canal. In 1845 a national Convention of Infidels was held in New York City.

Freethought had a long history in Germany, so when German immigrants arrived in the 1840s and 50s they also formed or joined institutions espousing freethought.

Draper Rides the Freethought Wave:
Evolution, Freethought, Anti-slavery, and the Protection of Children

Freethinkers were very much involved in the anti-slavery movement of the mid-1800s. They considered slavery to be a moral outrage and they were of course entirely happy to dismiss the idea that slavery was ordained by God, as many Southern preachers claimed.

Further, in the years prior to the Civil War, abolition was not just opposed by churches in the South, it was also *ignored* by churches in the north.

The only ministers who tended to embrace abolition in the north were radicals within liberal denominations such as the Quakers and the Unitarians.

There was therefore a great deal of anticlerical rhetoric among abolitionists, which led to abolitionists in general being labeled as "atheists." It is therefore ironic that today many people think that the abolition of slavery was somehow accomplished by churches.

Former slave and antislavery campaigner Frederick Douglass himself, having noticed a great deal of pro-slavery rationalization among church leaders in general, said:

> For my part, I would say, welcome infidelity! Welcome atheism! Welcome anything! in preference to the gospel, as preached by those Divines! They convert the very name of religion into an engine of tyranny, and barbarous cruelty, and serve to confirm more infidels, in this age, than all the infidel writings of Thomas Paine, Voltaire, and Bolingbroke, put together, have done.[187]

Meanwhile, prior to Darwin's publishing *Origin*, many religious people found it convenient to believe that God had *separately* created each of the human races. They further believed that the races were not therefore equal, and that the different races were practically different species. This made owning humans of another race not that much different from owning animal livestock, and it was a very convenient and comforting excuse for the many people who wanted slavery to continue.

Therefore, when Darwin came along and asserted that all human beings shared common ancestry, people wishing to continue slavery wanted and even needed to reject evolution. Darwin's theory simply destroyed the idea that God created each race of humans separately.

It is therefore clear why the science-loving, slavery-hating, religion-rejecting freethinkers seized on evolution, defended it, and promoted it.

It should also be noted that freethinkers also seized on evolution to help develop the first legislation for the protection of children. The Society for the Prevention of Cruelty to Animals pointed out that children (according to evolution) are members of the animal kingdom and should therefore at the very least get the protections that the law already gave to animals! That is how we originally came to have laws against child abuse in this country.

Abolition was not the freethought movement's only cause, and after the Civil War they continued to champion rational thought unimpeded by religious dogma, by the pronouncements of clergy, or by the assertions of holy books. They championed science as a source of knowledge. And once *Origin* had been published, they wholeheartedly gave their assistance to the cause of getting the public to accept evolution. They also helped many other causes, including those for women's rights, women's suffrage, birth control, and the rights of African Americans.

They were very influential from the end of the Civil War until roughly the time of U.S. entry into World War I, a period sometimes called the Golden Age of freethought. Public lectures were one of the primary forms of entertainment at the time, and the freethinkers had one of the best orators in the business: Robert G. Ingersoll

Ingersoll was the freethought movement's most famous proponent in the United States. He was a lawyer by profession and had served on the Union side during the Civil War. During, and especially after, the war, he built a reputation as a public speaker, which eventually made him famous. He packed the largest theaters of his day, charging the then-substantial amount of $1.00 per seat. Between 1865 and 1899 he traveled throughout the United States on more than a dozen speaking tours. It was literally the case that no human had been seen and heard directly by more Americans until the beginning of the radio, film, and television industries. Although he spoke on various subjects, including Shakespeare, the poet Robert Burns, the lives of famous patriots and scientists, and the political and moral issues of the day, his stance was always that of a freethinker. He was also a sought-after political speech maker for Republican presidential candidates, who later invited him to visit the White House. His speeches were humorous, entertaining, and generous. Clerics were often spotted in the crowd.

Ingersoll's reach and popularity were wide. In addition to presidents, Ingersoll was also friends with Mark Twain, Andrew Carnegie, Walt Whitman, General Ulysses S. Grant, Thomas Edison, and even the abolitionist preacher Henry Ward Beecher. His level of oratory was such that Mark Twain himself wrote of him: "What an organ is human speech when it is employed by a master!"

By now readers will have noticed the numerous points of commonality between Ingersoll and Draper. Although they were in different professions, they were both public figures. They both thought that science was the best way to obtain knowledge, and said so publicly. They both stated openly that they believed that religion suppressed scientific progress. Both publicly supported the acceptance of evolution by natural selection and put their professional public weight behind this. Both were esteemed public figures and both men's lives were centered in their later years in New York State. It is therefore clear why Ingersoll and rest of the science-loving, religion-rejecting, evolution-promoting freethinkers seized on John William Draper and lauded him, esteemed him, and promoted his work. It is undoubtedly the case that the men knew of each other, and each other's work, though there is no evidence that they ever met. Ingersoll's public speeches, pronouncements, and overall outlook may well have influenced Draper's, and we know that Draper's work was noticed and lauded by Ingersoll.

Draper's Conflict Thesis

Draper made his position on science versus religion clear in his 1874 book, the *History of the Conflict between Religion and Science*. The book largely examined the deliberate science-suppressing activities of the Roman Catholic Church, and even said some relatively nice things about Greek Orthodoxy, Islam, and Protestantism by way of comparison, but only when they did not hold back science. He pointed out that the Roman Catholic Church's ability to suppress science had largely been a result of its worldly power—so he left room for the idea that other religions could also suppress science if they were powerful and unified enough.

In fact, he duly pointed out that the real religious enemy of science is scriptural literalism, and that both Protestantism and Catholicism were guilty of this dangerous and repressive position:

> The two rival divisions of the Christian Church—Protes- tant and Catholic—were thus in accord on one point: to tolerate no science except such as they considered to be agreeable to the Scriptures.[188] [189]

He believed that science and Protestantism could in principle get along, and indeed that they "were begotten together and were born together." However, this assumed that Protestant churches should and would acknowledge science's superiority for discovering facts about the material world. Moreover, it assumed that when newly-discovered facts conflicted with old beliefs, it was the old beliefs that needed to be modified. In an oddly prescient passage, he criticized and bemoaned the fact that Evangelical Protestants were, already, making themselves enemies of science. Thus, although Draper aimed most of the criticism in his book at Roman Catholicism, his title made clear his opinion that religion and science themselves are in fundamental conflict.

This book was Draper's last major work, and his most controversial. It went through twenty editions in its first ten years and was translated into nine languages. It was reprinted as recently as 1998. Its ultimate accolade came when the Vatican banned the book, placing it on its *Li- borium Prohibitorium* (the index of forbidden books).

Draper's Evolution Lecture

On October 11, 1877, Draper gave a remarkable speech to Unitari- an ministers at the Unitarian Institute in Springfield, Massachusetts, titled "Evolution—its origin, progress, and consequences." This was published in the December issue of *Popular Science Monthly*.[190] In it he traced the history of science, the history of organisms, the develop- ment of the idea of evolution, and an explanation of evolution, along with comments about the suppression of science by religion that were, if anything, even more strongly worded than his previous comments on the topic. Islam did not escape criticism this time. He closed it by pleading with the ministers to not reject this all-important theory.

Writing

In addition to *Conflict*, Draper published on a wide variety of topics throughout his career. He continued to publish articles on chemistry and light. He wrote books on chemistry, human physiology, and natural philosophy. He wrote a three-volume *History of the American Civil War*. In addition, he never gave up the more speculative side of his nature, publishing *Thoughts on the Future Civil Policy of America* in 1865, and in 1862 a book that developed further the subject of the speech he had given in Oxford to the British Association: *A History of the Intellectual Development of Europe*.

It is clear that Draper, in addition to possessing a lively and curious mind, genuinely wanted to do good in the world and allow science and scientists to give their gifts to the world, free of persecution.

Draper's health began to fail in 1881 and, like most people of the time, he eventually died at home, where he lived on Hastings-on-Hudson. Interestingly enough, while his funeral service was in the East Village close to New York University, his remains are not interred in Manhattan, or Hastings-on-Hudson, or even Virginia. Nor was Draper's body returned to Great Britain as a favorite son come home to rest in peace. John William Draper lays comfortably at rest in Greenwood Cemetery in the Park Slope section of Brooklyn, New York.

Family Life and Legacy

As for home life, John and Antonia would have six children over the course of their marriage. Two of their children would live into the twentieth century, although one child, young William, would not live past eight, dying in 1853. William's death served as yet another emotional connection to Darwin, who lost the "apple of his eye," his Anne, to tuberculosis when she was just ten years old in 1854. The Drapers two eldest sons, John and Henry, did not have children. However John and Henry's three younger siblings, Virginia Draper (Maury), Daniel Draper, and Antonia Draper (Dixon) would each marry and give the elder Drapers a total of seven grandchildren.

Some of Draper's descendants continued to be interested in the interactions between chemistry, astronomy, light, and photography.

Draper's son Henry Draper (1837–1882) began his career as a physician. He graduated with his medical degree from NYU's medical school, an organization founded by his father. And like his father, he was a skilled doctor and also photographer of celestial bodies. He later married the wealthy Mary Anna Palmer, who helped him to advance his passion for astronomy and aided in some of his pioneering astronomical work. Among other projects, such as writing a chemistry textbook, he focused his photographic work on studying Venus and was the first to photograph the Orion Nebula. In fact, the "Draper Crater," located within the *Mare Imbrium* area on our moon is dedicated in his honor.

He also became a leader in the new field of photographing the faint spectra of stars—the rainbows of their light as spread out by a prism at the eye end of a telescope. These spectra, like that of the sun, showed many thin, dark lines apparently caused by various materials at various temperatures in a star's atmosphere. But most stellar spectra differed from that of the sun, in ways small and sometimes large.

His name lives on in the Henry Draper Catalogue of the spectra of a quarter million stars; it echoes in the news today every time a new planet is discovered orbiting a star named with its "HD" catalogue number. Henry died at age forty-five while just beginning to classify these in large numbers. His widow Mary Anna endowed several astronomical projects in his name, most famously the Henry Draper Catalogue, compiled at Harvard Observatory during subsequent decades.

Henry's brother John would also die just as he was perhaps entering his most productive years. John Christopher Draper died at fifty years of age from a bout of pneumonia in 1885. He was a surgeon and chemist who, like his elder brother, also attended NYU Medical School. He taught science and was an esteemed professor of the natural sciences and chemistry. He taught both in Europe and at New York University, specifically in the area of chemistry.

John William Draper's granddaughter Antonia Maury (1866–1952), daughter of Virginia Draper and the naturalist Reverend Mytton Maury, is well known today as one of the greats of early twentieth-century astronomy. Surrounded by scientific ideas since childhood, she graduated from Vassar College in 1887 with honors in the "hard sciences" of physics and astronomy—highly unusual for a woman in those days. That got her a job at Harvard College Observatory-—but only in low-level, routine toil as a "computer" (pencil-and-paper number cruncher) and, along with a roomful of other low-paid women, categorizer of vast numbers of the spectra of stars recorded on photographic plates. Undaunted, she paid close attention to patterns that became apparent among the tiny spectra she and the other women were laboriously examining and measuring with microscopes. She and some of the other women then worked out the basic classification system for stars that astronomers have used ever since. Maury also performed other pioneering astronomical work. Only late in life and posthumously were she and the other Harvard Observatory women granted the scientific recognition they had earned. The tale of their work is now standard in astronomy texts.

CHAPTER 11

ALFRED RUSSEL WALLACE

"Why did so many of the greatest intellects fail, while Darwin and myself hit upon the solution of this problem— a solution which...proves to have been (and still to be) a satisfying one to a large number of those best able to form a judgment on its merits?...

On a careful consideration, we find a curious series of correspondences, both in mind and in environment, which led Darwin and myself, alone among our contemporaries, to reach identically the same theory.

First (and most important, as I believe), in early life both Darwin and myself became ardent beetle-hunters. *Now there is certainly* no group of organisms that so impresses the collector by the almost infinite number of its specific forms, *the endless modifications of structure, shape, colour, and surface-markings that distinguish them from each other, and their innumerable adaptations to diverse environments.* [emphasis added]

Alfred Russel Wallace
(From Wallace's acceptance speech on receiving the Darwin-Wallace Medal in 1908, in The Darwin-Wallace Celebration Held on Thursday, 1st July 1908, by the Linnean Society of London, 1909)

"The Creator, if he exists, has an inordinate fondness for beetles."

Attributed to J. B. S. Haldane, from a speech to the British Interplanetary Society in 1951

Briefly:

Alfred Russel Wallace was a contemporary of Darwin. Wallace was also a British field biologist and naturalist. While Darwin was methodically researching and writing *Origin* for almost twenty years, Wallace was doing his own research and developed the same theory of natural selection. In fact, he had sent Darwin a manuscript copy to edit of a book he was planning to have published on the topic. This forced Darwin to move up his own publication of *Origin*. A key supporter of Darwin, Wallace wrote several scientific papers (1862, 1867, 1870) rebuking religious denial of evolution and natural selection. However, Wallace was also a strong proponent of phrenology, something that is fully rejected by both modern science and the discipline of anthropology. His ideas concerning spirituality and its impact on human development were equally controversial.

Where was Wallace?

Alfred Russel Wallace was not at the infamous 1860 meeting of British Association that turned into a circus. Nor was he at the 1858 meeting of the Linnean Society when his and Darwin's discovery of evolution by natural selection was first introduced to the scientific world.

He was, most likely, to the envy of scientists everywhere, out in the field doing his research. Darwin may have been staying out of the fray at Down House, but Wallace was even further removed. At the time of both those meetings, Wallace was not even in the same country—and barely in the same hemisphere—as the combatants.

Yet Wallace was the one who set that whole story in motion. He was the one who wrote an essay on natural selection that he sent to his admired correspondent Charles Darwin, asking him to send it on to the even-more esteemed Sir Charles Lyell. He was the one who therefore inadvertently struck a terror into Charles Darwin that was even greater than Darwin's terror of the backlash that would surely follow his announcement of the discovery of evolution by natural selection. That terror was, of course, the terror of not getting credit for the discovery.

Wallace was the one who therefore galvanized Hooker and Lyell into action, motivating them, at Darwin's behest, to announce Darwin and Wallace as co-discoverers of evolution, and read their papers as a joint offering at the Linnean Society meeting of July 1, 1858. Wallace's work was directly presented at that Linnean Society meeting, and of course published as a part of the proceedings of that meeting.

But Wallace was also the one who therefore galvanized Darwin himself into abandoning his work on a multi-volume proof beyond all possible doubt of evolution and into writing instead a readable and understandable volume describing his research, analysis, and discovery: *On the Origin of Species*. Darwin was motivated to write it quickly, too. The *Origin* came out in November 1859, only a year and nine months after Wallace first thought of natural selection, and only seventeen months after Darwin's June 1858 receipt of that fateful letter.

It is reasonable to say that without Wallace there would have been no *Origin*. Darwin might have published some book or books on evolution eventually, but they would not have been *Origin*. Or he might not have published at all. His health was poor and his capacity to do still more research on the subject seemingly endless. Darwin lived to be 73 years old, and he might have died before he published, leaving Hooker or Huxley with the unfortunate task of publishing his essay from 1844 and hoping for the best. Thus Wallace's work was indirectly present at the publication of *Origin*, and at all that ensued. He may not have had a hand in the writing of *Origin*, but he definitely had a foot in Darwin's backside that made him do it.

Wallace realized the importance of his discovery.

> The more I thought over it the more I became convinced that I had at length found the long-sought-for law of nature that solved the problem of the origin of species.191

But he had neither the interest nor the means, at that time, to mount a full-fledged campaign to successfully argue for the validity of his discovery against both the established religion of England—the Church of England—and some well-established scientific insiders like Owen. He had to work for a living, and lacked an academic post that would have provided a stable income. Thus, both by choice and by necessity, Wallace chose to continue to be abroad after he discovered evolution by natural selection, doing research and also making a living by getting specimens for those Victorian obsessives known as collectors.

He always stood by his evolutionary work and thrived on public debate, but he was personally shy and even self-effacing. Who was he, and how did he come to occupy this pivotal position in scientific history?

To say that Alfred Wallace had "skin in the game" regarding the theory of natural selection would minimize his research contributions, so much so that our understanding of how biological change occurs and how ecology impacts all living things and their evolution, would indeed leave much of the story of natural selection sparse. One can make the argument that the time was simply right for natural selection to be "uncovered" by naturalists since both Wallace and Darwin essentially came to the same conclusions studying very different species at different times in their careers.

But whatever your feelings about Wallace's place in the story of natural selection, we must conclude that their individual research would certainly make both Darwin and Wallace, who was fourteen years Darwin's junior, true men of their time and that of human history. As his modern advocates will tell you, Wallace should rightfully be lauded as the co-discoverer of evolution by means of natural selection.

Early Life and Career

Wallace's family was an unfortunate example of that classic Victorian archetype, the family that was once of means, that slowly slid into genteel poverty. This had consequences for Wallace's education, career, and financial security. Wallace was born in Usk, Monmouthshire, England (now a part of Wales) in 1823. He was the seventh of nine children. His father presided over a string of disastrous investments, and although trained as a lawyer, never practiced law.

In 1828, the family moved to Hertfordshire, England, where Wallace's mother had family. Wallace's only formal education came at the grammar school there. Then, around 1835, the family fell on even harder times, when Wallace's father was swindled and lost his remaining assets.

Wallace's story at this point resembles Huxley's. The family was too poor to educate him formally any further, but they had inculcated in him a love of learning. Having failed at the prosperous middle-class lifestyle of living off of their investments, the family had to send their son out to work at a relatively young age. However, the family was still middle class enough that ways were found for Alfred to learn ways of making a living that paid better than being a common laborer. On the other hand, living in the Welsh countryside and seeing the farmers who were struggling in true poverty made him forever conscious of the suffering and unfairness of the lives of the truly poor. At the age of fourteen, Wallace had to leave school and went to live with his older brother John in London, who was a carpenter.

Living in London enabled Wallace to take advantage of the centers for higher learning for the working class that were available there. At the Mechanics' Institute he heard lectures about the social reformer Robert Owen. These made him skeptical about the British class system, and his brother's progressive views plus reading the pamphlets of Thomas Paine made him skeptical of organized religion. Throughout his lifetime, Wallace was a huge proponent of social reform and class struggle, firmly committed to the welfare of the working class

Like all of Darwin's apostles, Wallace had a love for reading as a young man, and while he was getting his hands dirty plying his many trades he was also sharpening his mind. He read fervently and intensely, and in particular was captivated by books concerning social reform. Perhaps this is why for the rest of his life Wallace was considered an "out of the box" thinker.

Two years after moving to London, he left and moved in with his eldest brother, William, who was a land surveyor in Bedfordshire. Thus it was that Wallace, like the other apostles, found a profession, though he was unique among them in being trained as a surveyor rather than as a physician.

Surveying led to Wallace spending a great deal of time outside, wandering throughout the countryside, both in England, and later in Wales, where the brothers moved in 1841.[192] This led to an interest in the plants that he saw while surveying, and he both started a collection of pressed specimens, and bought books from which to learn plant identification. In this way Wallace and Huxley again resemble each other, teaching themselves, and gaining more knowledge however they could. In addition to botany, Wallace developed interests in geology and astronomy.

The surveying work also further developed Wallace's radical views. Surveying was in demand because of new laws that had been passed, that turned land that had previously been common land into land that was enclosed, and often privately held. While surveying, Wallace came into contact with the farmers who were suffering as a result of these new laws, and again he was moved toward a radical view of society, in opposition to the English class system. One of his earliest writing efforts was an essay called "The South Wales Farmer." Eventually there were violent uprisings in Wales, which led to an end to Wallace's surveying work in the area. As a result, Wallace became a school teacher for one year in 1844, at a boys' school in Leicestershire, called the Collegiate School in Leicester.

The year in Leicester was formative for him in many ways. The city had good libraries, and Wallace took advantage of them. He read several works on natural history and systematics—the science of how biological organisms are classified. Throughout his early adulthood, Wallace read on a wide variety of subjects, including history and political economy as well as natural history.

His readings included Darwin's *Voyage of the Beagle*, Lyell's *Principles of Geology*, Malthus's famous essay on his principle of population, Chambers' anonymously published *Vestiges*, and Humboldt's works on biology and exploration. *Vestiges*, however much it made other biologists cringe, introduced Wallace to the whole question of how species originate—the origin of species—and led to his reading further on the subject, including reading the works of Lamarck.

Wallace was also introduced to phrenology—the idea that the outer surface of the skull could be "read" and used to determine personality traits in the owner of the skull. In Leicester he also saw a stage demonstration of mesmerism, and felt that he was able to reproduce many of the mesmerist's effects—which led him to believe that it should not be dismissed as a simple trick. His interest in pseudoscientific mental and nonmaterial phenomena resurfaced at various times during his career.

While in Leicester, Wallace did two other crucial things. He met Henry Walter Bates, and following Bates' example, took up beetle collecting. Beetles are an enormous order of insects, with three hundred and fifty thousand described species in the world, more numerous than any other order in the animal kingdom. Even within the confines of England, about a thousand species of beetles could be found within ten miles of where Wallace lived. He became fascinated with studying beetles and beetle collecting. He remarked on "their many strange forms and often beautiful markings or colouring."

The sheer plenitude of beetle species available for collection and study, even to someone of Wallace's limited circumstances, made it possible for him to become experienced in identifying and collecting natural specimens. Collecting and studying beetles, with Bates, led to the next era in Wallace's life, but not immediately.

When William died in 1845, Alfred quit his teaching job, moved to Wales, and looked after his brother's business, as well as building a mechanics' institute in Wales with his brother John, and surveying for a proposed railway line. But running the business without William did not agree with Wallace. During this time, he kept up his correspondence with Bates, and read the book *A Voyage Up the River Amazon* by William H. Edwards.[193] This led to the idea that Wallace could make a better and more interesting living as a collector of exotic specimens, which would also allow him to do science.

Wallace became deeply enamored with the ideas of naturalist travel thanks to Bates. He was also inspired by reading the exploits of men like William Edwards, an American industrialist who studied the entomology and natural history of the Brazilian rainforest, and the Prussian naturalist Alexander von Humboldt, whose exploration and collecting work in Mexico, Venezuela, what was to become the nation of Ecuador but at the time was still under Spanish control, Cuba and the Andes mountains captivated Wallace.

It is at this juncture that Wallace's life again both resembles and is different from those of the other apostles. They all went on expeditions, with the exception of Draper. But all the others generally went with some level of help from government or other institutions, both for the project of the expedition itself, and for passage. Wallace, on the other hand, went as a private professional collector. He persuaded Bates to join him, and they went together on a collecting expedition to the Amazon.

In his memoir titled *My Life*, Wallace discussed why he and Darwin, out of all biologists everywhere and throughout the past, were the only two to discover evolution by natural selection. Among the crucial influences on both of them, Wallace included travel:

> Then, a little later (and with both of us almost accidentally) we became travellers, collectors, and observers, in some of the richest and most interesting portions of the earth; and we thus had forced upon our attention all the strange phenomena of local and geographical distribution, with the numerous problems to which they give rise. Thenceforward our interest in the great mystery of *how* species came into existence was intensified, and—again to use Darwin's expression—"haunted" us.[194]

Thus it was that Wallace set out, with a full complement of Victorian English desires and ideas going with him—radical reforming political ideas, skepticism about the beliefs espoused by the Church of England, the rise of both real science and semi-spiritual pseudoscience, the desire to make a fortune, the urge to explore, collect, and do science, and in Wallace's case, the active desire to solve the puzzle of the origin of species.

Journey to the Amazon

Wallace and Bates arrived at Pará, at the mouth of the Amazon, on May 28, 1848. (Pará is now called Belém). They started off working as a team, but had a disagreement after a few months and split up.

Henry Walter Bates and Batesian Mimicry and Evolution

At this point, Bates and Wallace's stories diverge, but Bates still had a lasting impact on the field of biology. Bates was a highly skilled entomologist as well as a collector, and wound up spending eleven years in the region around the Amazon, compared to Wallace's four. Eventually Bates's health declined due to malaria and he came back to England.

He sent his specimens on three different ships, to avoid the possibility of losing everything in one shipwreck. In so doing, he benefited from Wallace's unfortunate example which we'll discuss later in this chapter. Bates estimated that he had collected 14,712 different species, claiming (perhaps exaggeratedly) that 8,000 of them were new to science.

However, despite this accomplishment, his most lasting contribution was in the field of evolutionary biology. Bates discerned the type of biological mimicry that now bears his name—Batesian mimicry. Bates figured out that some organisms, for instance, some snakes, are toxic and dangerous, so predators avoid them. However, some other unrelated species (for instance other snakes) may evolve to look similar to these toxic species, and therefore the predators avoid them as well. This type of mimicry helps the reproductive success of the species doing the mimicking.

For example, coral snakes are highly venomous, and usually brightly colored, with distinctive rings of red, yellow or white, and black. Scarlet kingsnakes (*Lampropeltis elapsoides*) resemble eastern coral snakes (*Micrurus fulvius*), and sonoran mountain kingsnakes (*L. pyromelana*) resemble western coral snakes (*Micruroides euryxanthus*). The kingsnakes are harmless lookalikes of the toxic coral snakes, and predators avoid them in the areas where both kingsnakes and coral snakes exist together. The brightly-colored mimicking kingsnakes therefore gain a reproductive and therefore evolutionary benefit in those areas that they share with the brightly-colored toxic coral snakes.

Batesian mimicry is considered to be one of the most telling demonstrations of evolution by natural selection, with the selection in this case being done by predators who avoid organisms that bear a resemblance to dangerous ones. The more closely the mimic resembles the dangerous species, the greater will be its selective advantage.

In the tropics, Bates found mimicry in butterflies. He studied *Heliconius* butterflies, and found that toxic species did not have to avoid predators, because predators avoided them. They were slow moving, and yet were not eaten by birds. Meanwhile, harmless edible species that resembled them could also avoid predation. The protective coloration of the non-eaten individuals would then be passed on to their offspring.

Remarkably, Bates also found that among *Heliconius* butterflies, one could find chains of related butterfly species that slowly transitioned from one to the next, with similar but not identical coloration and markings. This is an excellent illustration of how biological varieties can eventually become separate species.

When Bates returned to England, he wrote the book *The Naturalist on the River Amazon*. His work impressed both Darwin and Hooker. He was also a convinced evolutionist, and actively worked to have evolution accepted.

Wallace in the Amazon and Rio Negro

Wallace spent most of his time near the Rio Negro and in the middle portion of the Amazon River itself.

His reading of *Vestiges* had intensely piqued his interest in the question of the origin of species. *Vestiges* was more pseudoscience than science, but we have already seen that Wallace had a soft spot for pseudoscience, and in this case it turned a self-educated lay person into someone passionately interested in biology. *Vestiges* had convincingly argued against both creationism and Lamarckism but had not really put forward a mechanism to replace them.

Wallace was convinced that evolution took place, but did not know how. However, he also knew that most people, including at the time most scientists, did not and would not accept evolution without ample demonstrations of its existence. In this regard Wallace was brilliant, in that he figured out ahead of time what kinds of evidence would serve as demonstrations that evolution takes place. He realized that he could look at individual phylogenies and map out how they extended into the environment, looking at how the spread of these groups was affected by the surrounding geography and geology, and how they were affected by local ecologies. He realized that two particular elements could help him:

First, how the spread of a species outward from its area of origin was either helped by geography or hindered by it, and

Second, how the local ecology in any given area influenced the adaptations that a species exhibited, regardless of how related species looked and were adapted.[195]

To this end, while in the Amazon region, he learned (or tried to learn) the region's birds, insects, primates, fish, plants, and physical geography. However, despite all this admirable and useful biological work, he was unable at that time to figure out the mechanism by which evolution works.

He also made a point of studying the languages and ways of life of local peoples with whom he worked and lived.[196]

Eventually, Wallace's health became poor, and in early 1852 he decided that it was time to go back to England. He went down the Rio Negro and the Amazon back to Pará. From this point onward, misfortune seemed to dog his every step on his return home. He reached Pará on the second of July, only to find out that his younger brother Herbert, who had been working in and around Pará, had died. He also found out that all the specimens that he had been dutifully sending down the Amazon to be sent to England for the last two years had been sitting on a dock far inland, at Barra do Rio Negro (Manaus). Thus, he wound up having to arrange to get all of them, and himself, back to England.

He arranged passage for himself and all his specimens on the brig *Helen*, only to have the ship catch fire and burn at sea twenty-six days into the voyage. The ship sank, taking Wallace's specimens and many of his notebooks with it. Thus he lost most of what he had done for the preceding two years.[197] Wallace and the ship's crew survived and spent ten days in leaky lifeboats in the Sargasso Sea.

After he and his companions were picked up by a cargo ship, that ship also nearly sank in a series of storms. The vessel was also both slow and old, so Wallace's voyage home took eighty days.

What Did Wallace Lose in the Shipwreck?

Wallace escaped with his life, a few notebooks, and very little else of a material nature out of the *Helen*'s sinking.

In his book *My Life*, Wallace wrote the following:

> I cannot attempt to describe my feelings and thoughts during these events. I was surprised to find myself very cool and collected. I hardly thought it possible we should escape and I remember thinking it almost foolish to save my watch and the little money I had at hand. However, after being in the boats some days I began to have more hope, and regretted not having saved some new shoes, cloth coat and trousers, hat, etc., which I might have done with a little trouble. My collections, however, were in the hold and were irretrievably lost. And now I began to think that almost all the

reward of my four years of privation and danger was lost. What I had hitherto sent home had little more than paid my expenses and what I had with me in the *Helen* I estimated would have realized about 500. But even all this might have gone with little regret had not by far the richest part of my own private collection gone also. All my private collection of insects and birds since I left Pará was with me, and comprised hundreds of new and beautiful species, which would have rendered (I had fondly hoped) my cabinet*, as far as regards American species, one of the finest in Europe.[198]

* This refers to a person's private collection, or "Cabinet of Curiosities."

What did Wallace's Amazon Trip Achieve?

For all his misfortunes, Wallace's trip to South America was not a complete bust.

To begin with, Wallace managed to grab a few of his notebooks before the *Helen* went down. A live parrot also seems to have survived.

He also had specimens that he had shipped home years earlier, before he left Pará for the interior and the Rio Negro. His agent in London had also insured the collection that Wallace had had with him on the *Helen*. It was not insured for what it was worth even in economic terms, with Wallace getting 200 pounds when he estimated it to be worth about 500. In terms of the damage to Wallace's scientific career, the loss was far greater. The specimens he collected could have been studied and further information gleaned from them for years into the future, had they survived.

However, the Amazonian trip was not a complete scientific disaster for Wallace. Remember—Wallace *started* his trip as a convinced evolutionist—he was just looking for the mechanism. While in South America, he had figured out that biogeographical distribution was likely to be a key to figuring out the mechanism of evolution. But hadn't gotten further than that.

He had, however, had an excellent apprenticeship in field biology and biogeography. He had seen the fantastic biological diversity of the rainforest. He had absorbed how plant species could be distributed, even when those species were not near to one another. As Asa Gray found as well, species separated by hundreds and even thousands of miles could be related to one another. Wallace also found that certain animal species, even though able to fly, would remain on their particular river bank.

Because Wallace was not an academic and held no teaching position at any British university, he had to survive on the sale of whatever salvageable collections he was able to save from his return from Brazil. He also lived on an insurance settlement caused by the loss of his vast collections. This was also a time when he would publish about his experiences and observation during his journey to South America and when he would fully establish a friendship with Charles Darwin, the two exchanging cordial and respectful ideas about the mechanics of the natural world for the rest of their lives.

The negative experiences of loss did not deter Wallace from continuing his interest in collecting species from around the globe. He was encouraged in fact since what species he was able to save, describe and write about from his journals gave him the opportunity to connect with many different naturalists at home and abroad. This gravitas suited Wallace's ego, especially since he was a poor boy now holding his respectful own with wealthy elites and their social and academic constituencies.

In England, he was by then a fairly well-respected travelling naturalist. During the two years he spent in England, he gave several presentations, wrote some articles for scientific publications, and also two books. One was titled *A Narrative of Travels on the Amazon and Rio Negro*. This was a travel book, telling of his Amazonian work and journey. The other was titled *Palm Trees of the Amazon and Their Uses*. As the title implies, this was a study both of the palm trees themselves and of the uses that peoples in the Amazon made of them. It was based in part on drawings that had been rescued from the fire on the *Helen*, though a professional illustrator was also used.

He had also made, using his drafting skills, excellent maps of the Uaupés River and Rio Negro. These were so good that they became the standard maps of those areas, and were used for many years to come. The map of the Rio Negro was published by the Royal Geographical Society, and the Uaupés map netted him backing from the society for his next voyage, to the Malay Archipelago.

He had learned the usefulness of traveling with various trade items such as fishhooks, axes, and mirrors. And despite having endured dysentery, malaria, torrential rains, and a fire at sea, and despite having decided to never travel again, after two years in England he was on his way to the Malay Archipelago, this time with backing from the Royal Geographical Society. This was the trip that would change his life, and scientific history.

Wallace, Darwin, and South America

Keep in mind at this juncture that the two co-discoverers of evolution by natural selection found themselves in different situations both economically and scientifically when they got back from South America. Both had gone with the question of the origin of species on their minds. Darwin, when travelling around South America, had looked for Lyell's predicted Centers of Creation, and had not found them. Wallace, inspired by *Vestiges*, had tried to figure out a mechanism for evolution and had not done so. He had figured out that the geographic distributions of species were a good avenue for researching this question.

When Darwin got back to England, in 1836, he still believed in the fixity of species. Wallace did not believe in that even when he left England the first time.

When Darwin got back to England, he did not even know that his soon-to-be famous finches were all finches. He found that out in 1837 after he gave them to the London Zoological Society, which then allowed John Gould, a responsible ornithologist, to analyze them.

But after that, in 1837, Darwin was on the trail. He drew his famous evolutionary tree diagram in a notebook, and also wrote in a notebook "one species does change into another"—*but he still had not figured out the mechanism.*

By 1838 Darwin had read Malthus's essay and had made the intellectual leap from human populations in industrial, agricultural, and undeveloped economies to a more general application to the natural world as a whole. By 1839, Darwin had had dinner with Augustin Pyramus de Candolle, a much older Swiss botanist. Darwin had studied de Candolle's work back when he had been a student at the University of Edinburgh. De Candolle had written about "nature's war," referring to competition in na-

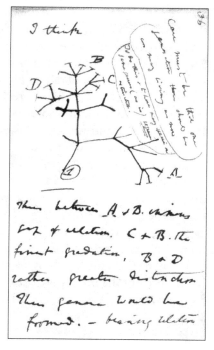

Darwin's 1837 sketch of an evolutionary "tree."

ture for space and resources. Again, this was another case where botanical work and ideas far outpaced those in zoology, again perhaps because people do not have the same emotional and personal reactions to plants and their predicaments as they do to animals—because despite most people at the time's not *wanting* to be related to animals, they nonetheless paid far more attention to those closer relatives than they did to the biology of plants. After having read Malthus, Darwin was able to relate Malthus's words back to what de Candolle had said much earlier. In 1838 or 9, he finally made the leap to evolution by natural selection, and suggested the idea to selected friends.

In 1839 he appears to have made a preliminary scribble on the subject, followed by an abstract in 1842, followed by his book-length essay in 1844. None of these were published.

Where the two men's careers were concerned, when Darwin returned to England in 1836, he was an established naturalist. The same was true of Wallace when he returned in 1852. Darwin wrote scientific papers next, and in 1839 published his *Journal of Remarks*, later known as *The Voyage of the Beagle*. Wallace wrote scientific articles and two less exciting books after returning from South America. Darwin came back with his collections intact, and his notebooks intact. Wallace had only a partial collection, and a few notebooks. They both came back having not yet figured out evolution by natural selection. However, Darwin could live on his independent income, stay home, and do research of a non-travelling nature. He could have his bird specimens analyzed by an authority on the subject. He could talk to pigeon breeders, and do experiments.

Darwin's health most likely precluded any further major journeys, so his next big things were more sedentary—and ambitious and voluminous—research and writing. This eventually included but was not limited to research and writing on the theory of evolution by natural selection, with the hoped-for end being a future enormous exposition on evolution by natural selection, one that would provide so much evidence and careful argumentation that it would quell all doubt for all time.

Wallace, on the other hand, needed to make a living. So Wallace's next big thing was his voyage to the Malay Archipelago, this time with some official backing. There he hoped to make his financial fortune, make major scientific discoveries, establish his scientific reputation, and figure out the mystery of the origin of species.[199] When Wallace left for the archipelago, he was hot on the trail of evolution by natural selection, but had not gotten there yet. Wallace and Darwin were both men of great ambition.

Wallace was a wanderer for intellectual and financial motivations for sure. In 1854, at thirty-one, he set sail again but this time in the opposite direction of South America, and well into Asia. Still a man without a family to support back home and the freedom such lack of commitment offered, Wallace spent the next eight years of his life travelling through present-day Malaysia and Indonesia collecting specimens and documenting his accounts of the natural world in this area of the globe.

The Malay Archipelago

The Malay Archipelago is the name given to the chain of tens of thousands of islands that stretch between the part of Indochina that is attached to the Asian mainland, and Australia. It divides the Indian Ocean from the Pacific. Most of it is now a part of either Malaysia or Indonesia.

Geologically, we now know it to have three separate parts. First, there is the Sunda Shelf, a stable continental shelf that is connected under water to the mainland of Southeast Asia. Although the connection is under water, it is not far under water—most of the platform is under less than 330 feet of water. It was at one time all above sea level. Second, there is the Sahul Shelf, a continental shelf connecting the island of New Guinea to the northern coast of Australia. It was once above water, and still has features showing where ancient streams eroded it. And third, there is the area between these two, which, like much of the entire region, is tectonically active.

The ancient geology of this area, and the land routes that organisms therefore took to various places within the archipelago in prehistoric times, played a huge role in the species distributions that Wallace encountered and figured out on his way to discovering evolution by natural selection. Part of Wallace's genius was that he not only figured out the biological past of the Malay region, he figured out its geological past as well, based in part on his biological findings.

Wallace started his work there when he arrived in Singapore on April 20, 1854. He spent his first three months in Singapore, generally learning about the east, collecting insects, and training his English teenaged assistant, Charles Allen.

Remarkably, he was already making notes and developing a plan for a book he intended to write called *The Organic Law of Change*. Make no mistake—Wallace was already planning on finding the mechanism for evolution, and writing about it.

The Sarawak Law

Wallace shifted his operations to the island of Borneo, arriving there on November 1, 1854. He did some collecting, and then spent the rainy season indoors, in the Sarawak region of the island. The break from collecting imposed on him in this way gave him time to write a paper titled "On the Law Which Has Regulated the Introduction of New Species." This has become known to history as the Sarawak Paper, which stated the Sarawak Law. It was an anti-creationist article, written in response to other papers then in circulation.

The law that Wallace asserted was as follows:

> Every species has come into existence coincident both in time and space with a pre-existing closely allied species.[200]

Species, he says, emerge by means of gradual change. As a result, he says that his law explains

> the natural system of arrangement of organic beings, their geographical distribution, the geological sequence, the phenomena of representative and substituted groups in all their modification, and the most singular peculiarities of anatomical structure.[201]

This was an excellent summary of existing knowledge. However, Wallace was able to add his own distinctive understanding to it. He had noted the patterns of distribution of many different organisms, and could provide specific evidence for his broad statement. From his own collecting activities, he could demonstrate the following:

> When a group (meaning a genus or class of organisms) is confined to one district, and is rich in species, it is almost invariably the case that the most closely allied species are found in the same locality or in closely adjoining localities, and that therefore the natural sequence of the species by affinity is also geographical.[202]

This allowed him to answer questions that had come up in his research/collecting activities, such as "Why are the closely-allied species of brown-backed trogons [trogons are a type of bird] all found in the east, and the green-backed in the west? Why are the macaws and cockatoos similarly restricted?"

In this paper, he even referred to Darwin's work, as reported in the *Voyage of the Beagle*. He said of the organisms populating these remote islands:

> They must have been first peopled (peopled in this case meaning colonized by plants and animals), like other newly formed islands, by the action of winds and currents, and at a period sufficiently remote to have had *the original species die out, and the modified prototypes only remain*. In the same way we can account for the separate islands having each their peculiar species. [203]

Wallace got so close. He *knew* that evolution takes place, not special creation or anything else. He stated it without hesitation, and he had the data to prove it! He stated that it took place through the gradual modification of existing species. Again, the only thing he was lacking was a mechanism.[204]

What Being a Professional Collector Did, Scientifically, for Wallace

Since Wallace had to earn money while in the archipelago, and in fact hoped to make his fortune there, this meant that on his numerous collecting forays he did not want to take just one individual of a given species as a type specimen, as a scientist might do. Instead, he wanted to collect a lot of specimens of everything, especially those that might excite collectors back in England. He would collect whole series of organisms, not just samples. What this meant was that he saw a great deal of variation of forms within a species.

In a way, the very of concept of a type specimen lends itself to essentialist thinking—that there is one ideal version of a species that is correct, and all variants are somehow less perfect examples of a species. When people believe that all species are perfect and specially created by God, intra-specific variation is a blow to the entire system of thinking.

Simply noticing the level of variation within a species that Wallace saw and demonstrated in his collections would tend to lead a person away from the essentialist idea that every species was fixed around an ideal type; and also tend to lead one away from the idea that each species was fashioned in an exact particular way by God.

Wallace started out in the Malay Archipelago already not believing in special creation, but seeing and recording and collecting all that intra-species variation—leading as well to some very blurry lines between species—may well have helped him on his road to the discovery of natural selection.

Reception in England of the Sarawak Paper

The Sarawak paper was published in September 1855 in the *Annals and Magazine of Natural History*. Before sending it, Wallace, already thinking of a larger exploration of this and related subjects, copied important passages from this paper into his species notebook.

Interestingly, the Sarawak paper caused at least one person to warn Darwin that he had a serious scientific rival. Lyell, originally a creationist whose own writings had suggested that species were created by God at specific "Centers of Creation," read the Sarawak paper and appreciated its significant implications. He started to think again about the idea of "transmutation," that is, the idea that one species can change into another. This, of course, is one important aspect of evolution, but one that Lyell had rejected until then. Lyell began a new notebook on species as a result. Then, when Lyell visited Darwin and Darwin finally let him in on his ideas about evolution, Lyell warned Darwin about the Sarawak paper. Darwin had read the paper, but had not fully appreciated its significance. At that point, Lyell also told Darwin to hurry up and publish his thoughts on evolution, because Wallace was coming near to having the same idea himself.

Darwin, in turn, had the decency to tell Wallace that Lyell had admired this paper, in a letter dated 22 December, 1857. He wrote:

> I am extremely glad to hear that you are attending to distribution in accordance with theoretical ideas. I am a firm believer, that without speculation there is no good & original observation. Few travellers have [at]tended to such points as you are now at work on; & indeed the whole subject of distribution of animals is dreadfully behind that of Plants.[205]

Note that Darwin is aware of how far ahead the field of botany is, regarding questions of species distribution compared to other fields of biology.

Darwin further wrote:

> You say that you have been somewhat surprised at no notice having been taken of your paper in the *Annals*: I cannot say that I am, for so very few naturalists care for anything

beyond the mere description of species. But you must not suppose that your paper has not been attended to: two very good men, Sir C. Lyell & Mr. E. Blyth at Calcutta specially called my attention to it. Though agreeing with you on your conclusion[s] in that paper, I believe I go much further than you; but it is too long a subject to enter on my speculative notions.[206]

Note here how Darwin tells Wallace that he (Darwin) is working on something very big, too big to discuss in the confines of that letter. That Big Thing is of course, evolution by natural selection. He is perhaps feeling Wallace out by saying "I believe I go much further than you." Is he hoping to engage Wallace in discussion, with an eye to letting him in on his ground-breaking idea of evolution by natural selection, as he had by then done with Hooker, Gray, Lyell, and others? This is possible. Or is he establishing precedence, making sure that Wallace knows that he thought of evolution by natural selection first, should that need arise? Both may be true. Darwin does not mention evolution in this letter, but it is clearly present in Darwin's mind when he is writing this letter to Wallace.

The Famous Letter to Darwin from Ternate

Ironically, this letter from Darwin was received by Wallace at about the time that Wallace was discovering evolution by natural selection for himself. On the eighth of January 1858, in less than a few weeks after Darwin wrote this letter, and most likely before Wallace received it, Wallace arrived on the island of Ternate. There, with the help of a wealthy Dutchman, he was able to rent a roomy and pleasant house. It was his home base for the next three years.

From his home base he set off and eventually arrived at Dodinga, a village on the island of Halmahera. He started collecting, but then he got malaria. He stayed on Dodinga for a little over a month, and was either sick or weak for most of that time. Unable to do much collecting and stuck in his hut, Wallace went back to theorizing. He understood and had asserted in his Sarawak paper that species changed over generations, with a given species "becoming changed either slowly or rapidly into another." But *why* this happened was something that he still did not know.

Then one day, even while he was shivering from effects of malaria, feeling so cold that he wrapped himself in blankets while living in the tropics and the temperature was eighty-eight degrees, he thought of Malthus.

Malthus, you will recall, stated that *human* populations were kept in check through disease, famine, warfare, and accidents. In other words, human populations did not grow larger than they were because many people died. Malthus stated that this was especially true among "savage" populations rather than among "civilized" ones. Wallace had noticed that animals bred in great abundance, and yet their populations did not overrun their environments. Why was that?

> [A]s animals usually breed much more rapidly than does mankind, the destruction every year from these causes [disease, famine, etc] must be enormous in order to keep down the numbers of each species, since they evidently do not increase regularly from year to year, as otherwise the world would long ago have been densely crowded with those that breed most quickly. Vaguely thinking over the enormous and constant destruction which this implied, it occurred to me to ask the question. Why do some die and some live? And the answer was clearly that on the whole the best fitted live. From the effects of disease the most healthy escaped; from enemies, the strongest, the swiftest, or the most cunning; from famine, the best hunters or those with the best digestion; and so on.[207]

From this he intuited:

> Then at once I seemed to see the whole effect of this, that when changes of land and sea, or of climate, or of food supply, or of enemies occurred—and we know that such changes have always been taking place—*in conjunction with the amount of individual variation that my experience as a collector had shown me to exist*, then all the changes necessary for the adaptation of the species to the changing conditions would be brought about.[208]

So Wallace had figured out that populations are genetically diverse, and that when changes in the environment take place (and changes are always taking place!) then the individuals who were best suited, through the random luck of genetic variability, to deal with these changes were the ones who would survive, and then breed. So then, the traits that had helped them to survive would carry on into future generations. Those future generations, if these particular changes continued to be in effect, would wind up being less and less like the original stock.

That evening, as soon as his immediate malarial fit had passed, Wallace wrote his ideas down. Then, over the next two days, he wrote out his complete essay. He titled it "On the Tendency of Varieties to depart indefinitely from the Original Type." He was aiming to send it off to Darwin by the next post. Somewhat confusingly, he headed the letter to Darwin "Ternate, February 1858" though in February, when he had written the essay, he had still been on Dodinga. This is most likely because his base of operations was in Ternate, so that was the best place for his correspondents to send their letters, should they want to write back to him. He got back to Ternate on March 1, and the mail boat was due on March 9.

He sent the essay to Darwin, along with a note saying that

> I hoped the idea would be as new to him as it was to me, and that it would supply the missing factor to explain the origin of species. I asked him, if he thought it sufficiently important, to show it to Sir Charles Lyell, who had thought so highly of my former paper.[209]

Wallace did not ask Darwin to have the essay published.

Reception of the Ternate Letter

Wallace's letter and essay struck like a bolt of lightning when Darwin received it on June 18, 1858. Wallace had it right, and Darwin knew it. He may well have reproached himself at this point for not taking Lyell's advice and—realizing that Wallace was hot on the trail of evolution by natural selection—publishing something quickly in order to establish precedence. However, it was fortunate that Lyell was both the person who had warned Darwin about the potential threat from Wallace and the person to whom Wallace wished that Darwin would send his essay. So Darwin sent the essay on to Lyell, along with a letter saying, "Your words have come true with a vengeance that I shd. be forestalled."

On the other hand, Darwin had to admit that Wallace had laid out the case for evolution by natural selection in an admirable fashion.

> I never saw a more striking coincidence, if Wallace had my M.S. sketch written out in 1842 he could not have made a better short abstract! Even his terms now stand as heads of my Chapters.[210]

Ever the gentleman, Darwin even offered to publish his rival's work. He wrote to Lyell: "Please return me the M.S. which he does not say he wishes me to publish; but I shall of course at once write & offer to send to any Journal."[211]

Lyell's cooler head came up with the idea of a joint presentation of both Darwin and Wallace's work. It may well be that Lyell swung into action so quickly for the simple reason that he had been anticipating this problem. He and perhaps Hooker as well may have already been thinking about what to do should this eventuality occur, given the distant early warning that Lyell had sensed when reading the Sarawak paper.

Hooker quickly became involved. The three corresponded. But Darwin, who was grieving the loss of his baby son Charles, who died on June 28, distanced himself from the decisions.

It should be noted that in some sense, Wallace was lucky in his choice of correspondent. Although Darwin's friends struggled hard to find a way to acknowledge both of evolution's co-discoverers while simultaneously giving Darwin precedence, at least the paper landed in friendly intellectual hands. Darwin, Lyell, and Hooker all recognized the genius of the paper; but Hooker and Lyell had been softened to the idea of evolution by natural selection through Darwin's patient persuasion, in Hooker's case, over a period of many years. Had Wallace written to, say, Richard Owen, it is entirely likely that the paper would never have been circulated, and that if Owen had bothered to write back to Wallace at all, it would have been to crush him under the weight of the disbelief of the British establishment.

Thus it came to be that Wallace and Darwin became known as the co-discoverers of evolution by natural selection. Their works were publicly presented, together. Lyell made the arrangements, and Joseph Hooker made the formal presentation of the theory, with writings from both men, at the July 1, 1858 meeting of the Linnean Society. The presentation was published in the Society's proceedings soon thereafter.

At the time, this bomb of an idea was so new that the audience did not know what to say. They were also perhaps a little intimidated that the presentation had been arranged by such senior and established scientists as Hooker and Lyell. However, even at the time, there was the sense that the religious establishment (including religious scientists) would soon mount a counterattack. This, sadly, would not be treated as ordinary science. This was something that would be attacked, not just disagreed with.

Hooker wrote about that meeting:

> [T]he interest excited was intense, but the subject was too novel and too ominous for the old school to enter the lists, before armoring. After the meeting it was talked over with bated breath: Lyell's approval, and perhaps in a small way mine, as his lieutenant in the affair, rather overawed the Fellows, who would otherwise have flown out against the doctrine.[212]

Wallace had no notice that this was happening until well after the fact, given the slowness of mail between England and the other side of the world. Darwin and Hooker, however, both wrote to him after the event. On October 8, having finally received notice of the event, Wallace wrote to his mother:

> I have received letters from Mr. Darwin and Dr. Hooker, two of the most eminent naturalists in England, which have highly gratified me. I sent Mr. Darwin an essay on a subject upon which he is now writing a great work. He showed it to Dr. Hooker and Sir Charles Lyell, who thought so highly of it that they had it read before the Linnean Society. This insures me the acquaintance of these eminent men on my return home.[213]

Thus it was that in the immediate aftermath of Darwin's receipt of Wallace's letter, and the ensuing presentation before the Linnean Society, that Wallace went back to collecting, which he had never stopped doing. Darwin finally got to work on a "short" version of his life's work, the book *On the Origin of Species*—with Lyell, Huxley, and Hooker urging him on.

Why Did Darwin and Wallace React to Their Discoveries in Such Different Ways?

The cracking of a scientific mystery is usually a cause of joy in the heart of the discoverer. The pure pleasure of success in a difficult pursuit, coupled with a vision of future eminence, is enough to drive most people to publish their findings quickly. Yet Wallace's and Darwin's reactions to their discovery could not have been more different. Why did Darwin delay while Wallace sent his idea off in the next post?

Remember—where Darwin hesitated about publishing about his discovery of evolution by natural selection for twenty years, Wallace went ahead and submitted his theory to a colleague for review as soon as he could. He did not have the mountains of data that Darwin had, but he felt that he had enough to make his case. Wallace, like Darwin, also planned on writing a bigger book at some later date. But perhaps he had an inkling that Darwin was preparing something the same or similar, and Wallace did not want to be pipped at the post. Perhaps Wallace was the one who feared being an also-ran if he did not circulate his idea quickly.

Further, Darwin was clearly concerned for the social and religious blowback that would undoubtedly take place when the non-scientific public understood what his theory implied for the Church of England, for the aristocracy, and for the English class system in general. Darwin was ensconced in a pleasant country life in which he was on good terms with his neighbors, including the local clergy, and with his wife. Darwin's wife, Emma, knew about Darwin's ideas and knew that he had slowly become an agnostic. Although raised in a religiously free-thinking family, Emma was nonetheless concerned that she would go to heaven and her beloved Charles would not.

Wallace, on the other hand, had no such compunctions. He was already disenchanted with the Church of England, the aristocracy, and the English class system in general. He knew that his family felt the same. Further, he was far away in the Malay Archipelago and immediate English social blowback would not affect him at all. On top of that, he needed to establish himself as a scientist. If he could return to England as a sort of conquering scientific hero, the man who had finally cracked the code of the origin of species, it might do wonders for his professional and financial stability. His letter to his mother shows how pleased he was, that he would now be able, when he went home, to make the acquaintance of important scientists like Lyell, Darwin, and Hooker. And on top of that, Wallace described himself as "a young man in a hurry." So Wallace was going to put his ideas out and let the social chips land where they may.

Darwin was not a man in a hurry, and it took Wallace's actual willingness to go public with his big idea that finally got Darwin to go public with his. The rest is history.

Biogeography

Why do regions that have similar climates have different plants and animals? This was a mystery that had long stumped European biologists. In our modern world, we take it for granted that different plants and animals live in different places. But it came as a surprise to Europeans when their sixteenth century explorers came back and described animals on other continents that were very different from the ones found at home in Europe.

Why should there be different animals in different places, if their climates were the same? If God created plants and animals to be exactly suited to their environments, then why were there different plants and animals in different places, even when the environments were similar? Why did explorers bring back hummingbirds, for instance, when none existed in Europe? Eventually, European naturalists decided that God had performed not just one act of original creation of all plants and animals, but six, scattered throughout the world, and for whatever reason he created wildly different plants and animals in these different places.

This works pretty well as an explanation until you look at actual species distributions. Wallace noticed, for instance, that mountain ranges and rivers often form a border to the range of a given species. Why should that be, when there was the exact same climate on the other side of the river? Six separate acts of divine creation don't account for that. The idea that each species is perfectly created for its environment does not account for that.

What does account for it is evolution. The geographical distribution of a species is simply a record of its inheritance. The study of the geographical distribution of species is the field called biogeography, and Wallace even more than Darwin, is the person who recognized it and developed it as a field.

Here's how this works: As a species spreads out, it colonizes new areas, with new challenges. So adaptations may start to take place in the colonizing part of the population. However, the species can only spread out in areas where it is capable of spreading. So, for instance, for many species, a wide river would be an adequate barrier to its movement. As a result, the species might continue to spread along the river bank, but would not spread to the other side. Then, if the range of this continually spreading species is divided, by the slow rise of a mountain range, a flood creating a new river, or a tectonic shift that submerges some of the land, for instance, the one broadly-dispersed population is divided into two or more isolated smaller populations, each with the characteristics that are peculiar to that area alone.

To repeat from earlier in this chapter because it is so critically important to note, what Wallace said about the Galapagos islands, in the Sarawak paper ("They" in this case refers to the islands themselves):

They must have been first peopled (peopled in this case meaning colonized by plants and animals), like other newly formed islands, by the action of winds and currents, and at a period sufficiently remote to have had *the original species die out, and the modified prototypes only remain.* In the same way we can account for the separate islands having each their peculiar species."[214]

Thus, evolution and biogeography are inextricably mixed. To fully understand one, you must understand the other. To understand where a species originated, you need to know its evolutionary history, and you need to know the geological history of the region where the species came from as well. What's more, as Wallace figured out, sometimes you can figure out the geological history of an area by looking at its species distributions. The idea that you could figure out the geological history of an area by looking at its living biological species was a profound and radical leap of intelligence.

Wallace's Line

Wallace's profoundest insight into the interaction between biology and geology came about while he was in the Malay Archipelago. Just as Joseph Hooker was able to use evolutionary theory to see into the deep past of the Cedars of Lebanon, and Asa Gray found that evolutionary theory—plus an Ice Age—provided the only satisfactory explanation of the disjunction that bears his name, just so, Alfred Russel Wallace was able to see into the deep geological past in a corner of the world far from his home, just by using evolutionary theory, and his own inexhaustible powers of biological observation. But in the case of Hooker and Gray, they were helped along in their insights by Darwin's careful tuition. Wallace figured it out all by himself.

In 1856, Wallace made the short journey from the island of Bali to the island of Lombok. These islands are about 15 ½ miles apart. Yet, when Wallace got to Lombok, he found a very different set of animals from the ones he had seen on Bali. On Lombok, he found Australian cockatoos and Australian birds called honeyeaters. Yet just across the strait on Bali, *and* on Borneo, *and* in Malaysia he had seen Asian birds such as woodpeckers, fruit thrushes, and oriental barbets.

Why was this the case? The islands were very near to each other. Borneo is much farther from Bali than Bali is from Lombok. Yet Borneo and Bali have animals that are similar to one another, while Bali and Lombok have very different ones. The climates on these islands are very similar, yet the animals on them are different.

He emphasized this in later writings, saying that the faunas of Bali and Lombok were more different from one another than the fauna of Japan is compared to that of Great Britain!

Why was this so?

Wallace wrote what he thought was the explanation:

> [T]he western part to be a separated portion of continental Asia, the eastern the fragmentary prolongation of a former pacific continent."

In other words, the island of Bali, which is west of Lombok, was the last tattered extension of the Asian continent, while Lombok just to the east was the last tattered extension of a continent that had included Australia. He was right about the Asian continent, and almost right about the Pacific one.

Now recall the geological regions of this area that were introduced earlier in this chapter. What we now know is that the Southeast Asian mainland *did* connect, on dry land, all the way to Bali. The Sunda Shelf is a stable continental shelf that was once *all above sea level*. It is connected to the Southeast Asian mainland. Bali was at one time the furthest-most dry land extension of that Southeast Asian continent. As a result, plants and animals from Southeast Asia were the ones who pushed out and lived on that nice dry continental shelf, all the way to Bali.

Then, that continental shelf slowly submerged, due to ice melting following an Ice Age. Islands like Borneo and Bali were the small bits of land that remained above sea level. This means that those Asian species were the starter populations for all the animals that lived on and evolved on those islands.

Meanwhile, there was also a continent that included Australia, plus what is now the island of New Guinea, and a great deal of what was then dry land in between. This was the Sahul continent. So species that were extant in Australia extended all the way out to New Guinea.

At the time, there were also scattered islands in between the Sahul continent and the Sunda one. The continental land mass of the Sahul (Australian) continent provided the jumping-off point for the species that came to inhabit that scattering of islands in between those two ancient continents.

Again, as sea levels rose, parts of the Sahul continent were submerged. Today, the parts of the former Sahul continent that are underwater are known as the Sahul shelf. The parts above water are the continent of Australia, and the island of new Guinea. Meanwhile, the scattered islands (including Lombok) between Bali and New Guinea make up the area that we now call Wallacea, named after Wallace.

And this is how it came to be that islands that were only 15½ miles apart, that had very *similar* climates and terrains, had very *different* animals living on them. And Wallace figured that out, just by looking at the biology.

Keep in mind that Wallace was living in a world in which most Europeans still thought in terms of a weird combination of the Great Chain of Being, Genesis, and six separate divine acts of creation. That he managed to figure out that Balian animals were evolved from Asian ones, while the Lombokian ones were evolved from Australian ones, and that parts of what had been continents had become submerged, all when the concept of an Ice Age was still new, and plate tectonics had yet to be discovered, is an amazing leap of intellect.

As he travelled further throughout the archipelago, he traced a distinct boundary snaking between the islands in this particular region. The boundary separates the different faunas, the different biogeographical regions, the islands containing species that trace their ancestry back to Asia, versus the islands containing species that trace their ancestry back to Australia.

He published this finding in his 1859 essay "On the Zoological Geography of the Malay Archipelago." Huxley then named this boundary "Wallace's Line."

Since Wallace first published this insight, adjustments have been made to the line, and other biogeographical lines have been added both in the Malay region and elsewhere around the world. But that basic division is still known as Wallace's Line (or the Wallace Line) in honor of the man who first figured out its existence, and its significance.

Wallace's Immediate Post-Malay Life

After the publication of *Origin* in 1859, Wallace served as a champion of Darwin when he returned from his Asia travels in 1862. Although Wallace was late to the game regarding the immediate reaction to Darwin's preeminent work, nonetheless he was a tireless champion for natural selection. As we will see in the next chapter, his numerous writings and speeches help cement Darwin's (and his) work regarding species adaptation. For a man without formal training or earned degree he was indeed swimming with many really big intellectual fish, and he never backed down from speaking the truth. This truth was and is that evolution by means of natural selection is the best and perhaps most obvious way to view our and every species' existence, life, change, and eventual demise.

When Wallace returned to England in 1862, he had collected over 125,000 plant and animal specimens, including almost 5,000 species of bird and insects that were new to science.

These included the remarkable and beautiful butterflies called Wallace's golden birdwing butterfly (*Ornithoptera croesus*) from the island of Bacan, and Rajah Brooke's birdwing butterfly (*Trogonoptera brookiana*) from the island of Borneo and named after Sir James Brooke, the "White Rajah" of Sarawak, an Englishman who ruled over the Sarawak region on Borneo, and who had invited Wallace to collect there; and Wallace's standard-wing bird of paradise (*Semioptera wallacei*), again from the island of Bacan.

Not long after his return he met with Huxley, Darwin, and Lyell and re-met his old friend Bates, who was finally back from the Amazon.

In addition to the Sarawak paper and his essay on natural selection, he had already published two other articles many other essays and two important articles from his Malay journey: an article on orangutans, and an article on the natural history of the Aru Islands, a group of islands in what is now eastern Indonesia.

At this point in his life, he wanted to sell much of his Malay collection, and use the proceeds to enable him to have a nest egg and settle down. The rest of his collection would be used for his own personal scientific research and writing. He wanted to get a job, marry, and do science.

But first he needed to know what was in his collections! From 1862-1865, Wallace worked to understand, them, biologically. During that period, he made at least sixteen presentations to his fellow scientists at various professional meetings. In doing so he was able to meet nearly every important English naturalist, and make friends with some of them.

Upon his return from Asia, Wallace sought to settle down with his 1864 courtship of a young woman who he would not name and who has been only known to history by the cryptic moniker of "Ms. L." However, the woman would later reject Wallace's affections. This unrequited love affair devastated Wallace, who was ready to start a family and live the life of a married man.

Wallace again attempted the semblance of a Victorian settled life by asking Annie Mitten, the daughter of colleague and fellow naturalist William Mitten, for her hand in marriage. They married in 1866 and their marriage produced two sons and a daughter. As with all the apostles who would have children, Wallace would lose a child to illness. Their first son, Herbert Spencer Wallace, would die at the age of seven in 1874,[215] however the other children would live well into the twentieth century. Wallace's second child, a girl who they named Violet Isabel, did not have any children and passed away at the age of seventy-six in 1945.

The Wallace's last child and second son, William Greenell, had three children and thus kept the Wallace line move into the twentieth century. Their progeny included two sons and a daughter. Sadly, their daughter Elizabeth would die within a month of her birth. However, William's sons Richard and Alfred would marry and have children of their own.

Like his father before him, Alfred Russel Wallace was a terrible custodian of his family's finances. He made consistently poor financial decisions throughout his adult life, investing in mines and transportation that saw no profit and left his family in a precariously bad financial position. His brother-in-law had a photographic business in which Wallace invested, and which wound up costing him about £700. Although this was not all of the money that Wallace made from his collecting activities, it nonetheless made for an unfortunate decrease in the nest egg that Wallace had hoped to have for his own future.

This precarious and continual loss of income was in spite of the ongoing sales of his books and the work he did to support his family by doing ad hoc clerical work for the British government. A journeyman since the time his parents pulled him from school as a child and later a survey worker in his brother's businesses, Wallace would never really work in a full-time position to earn a consistent annual salary. But fortunately his friends and fellow naturalists, including Charles Darwin, helped Wallace support himself and his family by offering him supplemental editorial work on their own publications.

Wallace and Victorian Science

Wallace possessed a fearless intellect. In Victorian England, this could be both a blessing and a curse. Recall that Wallace, when he figured out that natural selection is the mechanism by which evolution works, simply sat down and wrote up a cogent, evidence-filled essay on the subject and mailed it off to Darwin (to give to Lyell) without a backward glance. Darwin, by contrast, hesitated about going public with the idea for nearly twenty years. This fearlessness meant that Wallace got to share in the credit for the discovery of evolution by natural selection. If Wallace had kept his mouth shut, Darwin would most likely have eventually published about evolution on his own, if his health had held together. So Wallace's fearlessness wound up putting him on center stage in Victorian biology.

But it also meant that he did not play the careful game that Darwin and Hooker did. Wallace's family was already populated by radical thinkers who did not like the class system, the economic system, or the Church of England. So Wallace did not fear blowback from his immediate family when it came to stating and publishing radical ideas. Hooker himself complained to Darwin:

> It is all very well for Wallace to wonder at scientific men being afraid of saying what they think—he has all 'the freedom of motion in vacuo' in one sense. Had he as many kind and good relations as I have, who would be grieved and pained to hear me say what I think, and had he children who would be placed in predicaments most detrimental to children's minds by such avowals on my part, he would not wonder so much.[216]

So here we have the irony of one intellectual radical complaining to another intellectual radical about a third eminent biologist who was too radical for them both! Of course Hooker had an excellent job through the government that enabled him to do science and have financial stability at the same time. He was bravely willing to support evolution, but not willing to make his family socially uncomfortable. Darwin had independent financial stability, but also enjoyed his country life as a part of established, comfortable society, and had a family whom he did not wish to offend. But Wallace had "the freedom of motion in vacuo"—the freedom of a molecule in a vacuum to move around without anything else getting in the way. This freedom was an intellectual strength, and a social weakness. It also enabled him to embrace some fringe ideas, some of which were not intellectually sound.

Wallace's Work as an Apostle

In the years that followed Wallace's return to England, Wallace came to be known as one of Darwin's (and evolution's) "right-hand men."

Darwin appreciated Wallace's willingness to speak out boldly on the subject, and deliberately made a point of crediting Wallace with the co-discovery of evolution by natural selection.

Wallace cheerfully shredded a paper that claimed that the hexagonal cells found in bees' honeycombs could not possibly have arisen through natural selection. He wrote papers on Malay butterflies, bird's nests, bird's plumage, and direct commentaries on the origin of species controversy. In 1869 he published the influential and popular book *The Malay Archipelago* and in 1870, *Contributions to the Theory of Natural Selection*.

His works included writing on biogeography, natural selection, and glaciology. He wrote over 150 contributions in the next decade, including three books: in 1876 he published *The Geographical Distribution of Animals*; in 1878 *Tropical Nature, and Other Essays*; and in 1880 *Island Life*. All of these books are still consulted today. In 1880 he further cemented his support for evolution by natural selection by publishing a book titled simply *Darwinism*.

He did all this while raising a family, and participating in the Victorian equivalent of the gig economy.

Wallace's Other Activities

At the same time that he was advancing the cause of acceptance of evolution by natural selection, Wallace's fearless mind also took up other causes.

Spiritualism

In the 1860s, Wallace became interested in spiritualism. He took great solace in it after the 1874 with the sad death of his small child Bertie. He also felt that matters of the spirit applied to humans in other ways as well. Although he continued to accept and promote the idea that humans are animals and had evolved from common ancestors, he came to believe that human beings' moral dimension was divinely inspired—not by the Christian God, but by some other spiritual force. He published approximately a hundred writings on spiritualism.

This was probably a step too far for the British establishment. One could be an evolutionist and still be respectable, or a spiritualist and still be respectable, but not both. The scientists who were working to have evolution accepted did not feel that they were being helped by having Wallace—the co-discoverer of evolution!—carrying on in this way. They were working hard at having evolution accepted and did not want people rocking the boat with unnecessary side issues. However, a reasonable interpretation of Wallace's stance is this: Wallace, having long ago rejected the teachings of the Church of England, still hankered after the existence of some sort of spiritual realm. This was not so very different from those who accepted evolution while quietly continuing to believe in Christianity. Catholic thinkers in fact came up with a similar arrangement. If Wallace had quietly believed in some of the less rational teachings of the Church of England, this would have been no less irrational, but much more socially acceptable.

Anti-Eugenics

The term "eugenics" was coined after Darwin died but not before Wallace did. Wallace said the following on the subject:

> Why, never by word or deed have I given the slightest countenance to eugenics. Segregation of the unfit, indeed! It is a mere excuse for establishing a medical tyranny. And we have enough of this kind of tyranny already. Even now, the lunacy laws give dangerous powers to the medical fraternity. At the present moment, there are some perfectly sane people incarcerated in lunatic asylums simply for believing in spiritualism. The world does not want the eugenist to set it straight. Give the people good conditions, improve their environment, and all will tend towards the highest type. Eugenics is simply the meddlesome interference of an arrogant, scientific priestcraft.[217]

With the discovery of evolution by natural selection, a class-based society like England's had a problem. If society accepted evolution, it had to reject Genesis, and the whole Great Chain of Being idea, which had enabled the upper classes to justify their privileged position over many generations. Evolution, which says that all human beings descend from common ancestors, destroyed the idea that God had put aristocrats, and the wealthy, at the top of the mortal chain of being. So what was a privileged member of the English class system to do? Why, invent eugenics, of course. Eugenics was a convenient way for powerful and privileged people to again justify all their extra privileges while attempting to breed out unflattering human features or behaviors, usually concerning people who looked very different then themselves, while also believing they are at the top of the eugenics food chain. If God had not made them special, then by golly, evolution had! They felt that they were just biologically superior and this therefore justified their giving themselves still further advantages. Wallace was having none of it.

Women's Suffrage, Land Reform, and Other Radical Ideas

Wallace rounded out his radical ideas with full support of women's suffrage, land reform, and suspicion about smallpox vaccines. For women's suffrage, he made it plain, in an article that was published in the *Times* of London in 1909. It read:

> At a meeting in support of woman suffrage at Godalming last night, at which Sir William Chance presided, a letter was read from Dr. Alfred Russel Wallace, O.M.

> Dr. Wallace wrote:—'As long as I have thought or written at all on politics, I have been in favour of woman suffrage. None of the arguments for or against have any weight with me, except the broad one, which may be thus stated:—All the human inhabitants of any one country should have equal rights and liberties before the law; women are human beings; therefore they should have votes as well as men. It matters not to me whether ten millions or only ten claim it—the right and the liberty should exist, even if they do not use it. The term 'Liberal' does not apply to those who refuse this natural and indefeasible right. *Fiat justitia, ruat cœlum.*'[218]

Where land reform is concerned, he argued in favor of land nationalization. He recognized that the vast majority of the land in England was owned by only a few people. Since land was at that time a major avenue for being able to grow food and/or make money, Wallace opined that having most land being concentrated in the hands of a very few was unjust. He argued for the nationalization of land, in which the government would buy most of the land, and then rent it out to people who would actually use it.

His suspicions surrounding the smallpox vaccine were based on a critical reading of statistics, plus a dislike of government coercion, which he saw as mainly being aimed at the poor. The social justice issue was important. The Centers for Disease Control and Prevention reports:

> The Victorian vaccination legislation was part of an unfair, thoroughly class-based, coercive, and disciplinary health-

care and justice system: poor, working-class persons were subjected to the full force of the law while better-off persons were provided with safer vaccines and could easily avoid punishment if they did not comply.

The vaccines in use at the time really could be dangerous, and different ones were in use in different places. Wallace also argued that the statistics used to prove their efficacy had not been properly analyzed. For example, in the statistics used to justify mandatory vaccination, it had been shown that unvaccinated people had been more likely to die of smallpox. However, Wallace argued that these unvaccinated people had largely been living in squalor, and that was likely to account for their increased mortality. The vast quantities of statistics available to us today were simply not available in Victorian times, and Wallace was among the first to try to resolve a question in epidemiology by using statistical arguments. Wallace's stance may largely have been one against coercion, but it was also based on a knowledge of how poor people actually lived.

All in all, Wallace's sympathies lay squarely with the underprivileged in his country. Although he got along with aristocrats like Lyell, and members of the moneyed middle class like Darwin, his instinctive sympathies went with those who did not have the power and access to representation that they should have had.

Lack of Respectability and Lack of Money

Lack of money and lack of respectability kept reinforcing one another in Wallace's life. Victorian England was not generous in its jobs for zoologists or field biologists, and there were plenty who wanted them. Meanwhile, Wallace's fearless intellect and passion for social justice meant that he said and did many things that made people with power and money uncomfortable. Meanwhile, most of the money that he had made from selling his Malay collections had been poorly invested and the money had been lost. This meant that he needed a steady job and could not get one. Thus, he participated in the Victorian version of the gig economy, grading papers for government exams, lecturing, editing other people's books, and so on, in order to earn a little money. He also moved ever farther from London, as a way of saving money while having a pleasant place to live.

In another odd echo of today's world, in Victorian England, a country whose empire spanned the globe, and whose ships navigated that globe with ease, using spherical maps of the earth to succeed in their navigation, there was a noisy contingent of people who believed in a flat earth. One of them was wealthy enough to offer a prize of £500 to anyone who could prove that the earth is not flat. Wallace needed the money, so he took up the challenge and successfully demonstrated the curvature of the earth in the "Bedford Canal experiment". Not surprisingly, the flat earther was not willing to give up his money, and into the bargain he harassed Wallace for years afterward. This type of public notice, in turn, made Wallace even less employable.

Darwin was aware of Wallace's financial straits. He figured out that although he would not be able to get Wallace secure employment, he might be able to arrange a government pension for Wallace, in honor of his contributions to science. Darwin got numerous eminent scientists to sign a letter petitioning the government to award Wallace a pension. It worked. In 1881, Wallace was awarded a lifetime pension from the British government for £200 per year. It was even backdated to 1880. Wallace received news of this on his fifty-eighth birthday. Although this money alone was not enough to live on, it was a trustworthy and comforting safety net, which helped Wallace for his remaining thirty-three years.

Wallace as the Grand Old Man of Science (and Other Things)

Wallace never stopped being interesting. In addition to land reform, votes for women, eugenics, and vaccinations, in his later years Wallace interested himself in the labor movement, and equality of opportunity for all people. He wrote about the advantages of moving to paper money. He suggested that manufactured goods should actually be labeled as to their contents. In addition to being an invaluable resource for biogeography and evolutionary study, Wallace's book *The Malay Archipelago* also influenced the writings of the novelist Joseph Conrad. In 1907 Wallace debunked Percival Lowell's hot new idea that Mars was inhabited. He came close to realizing and stating the anthropic principle. (This principle is a response to the idea that it is some kind of very unlikely miracle that we find ourselves in a universe with the materials necessary for life, and on a planet so perfectly suited for our existence. The anthropic principle states that it is no miracle. The only place where life could evolve to the point of having people to wonder about it all is by definition a planet that is suited for life, no matter common or rare such planets may be.) He kept doing science. He interested himself in glaciation, and exobiology, as well as evolution and biogeography. He went on "botanizing excursions" to Wales, the English Lake Country, and Switzerland.

And he lectured. He attended meetings and lectured throughout England, and sometimes went to Scotland and Ireland. He addressed a group at Davos, Switzerland. His most ambitious post-Malay journey was a lecturing trip to the United States. Like Agassiz before him, Wallace was invited to give a series of lectures at the Lowell Institute in Massachusetts. Unlike Agassiz, he lectured on Darwinism. Once he had delivered the lectures, he was free to move about as he pleased, including lecturing elsewhere. In 1886 and 1887, he visited in and around Boston, Washington and New York, and met numerous important people, including the president of the United States, Grover Cleveland. Later in 1887 he travelled across the United States all the way to California. There, he was able to reunite with his older brother John, whom he had not seen for nearly forty years. He gave more presentations. He got to tour the site that would become Stanford University with Leland Stanford himself, toured Yosemite Valley, and visited redwood groves in the company of John Muir. In July of 1887 he returned to the east coast, and then back to England. He wrote about the topics that he had covered in his American lectures, and this resulted in one of his most frequently-cited works, the book titled *Darwinism*.

Wallace's final scientific work, *Man's Place in the Universe*, which postulated wrongly that only earth could support water-based organic life in the cosmos, was published in 1904. His autobiography published in 1905 and entitled *My Life* rounded out Wallace's significant life-long literary career.

To Wallace's credit he rejected the idea and concept of eugenics, breeding humans to their "fullest potential," seeing it as dangerous and allowing for its faulty science and methodologies to serve the goals and objectives of both scientists and lay people who wish to view Caucasians as superior to all other racial groups. However, he was a believer in and supporter of phrenology and an advocate for spirituality to explain the natural world (to Darwin's dismay). But this just may be because Wallace was a man of his times in other venues as well, not just when it came to developing or co-discovering natural selection. Certainly during the reign of Queen Victoria, spirituality and otherworldliness had captivated British society and to be *in* with the elite who believed in and accepted the occult in many ways afforded some level of social validation.

To Wallace's discredit, he did not believe people should receive vaccinations. His ideas were based on two fundamental views, each based on the concept of non-intervention. The first was based on civil libertarian ideals, Wallace felt the government should not dictate to its citizens what illnesses they should be inoculated against. Strange since he was such a supporter of the poor and disenfranchised who clearly have less access to medical care. The second reason was his idea that nature balances itself out and that transmittable disease should work freely on every population. In this case, nature should not be tampered with through the human ability to cure disease. In both cases, the modern anti-vaccination community would certainly respect his ideas, but they also do lead to serious issues regarding public health and respect for the lives of others.

As a man who understood and respected nature, Wallace certainly believed in ecological conservation. Perhaps it is because he was a "boots on the ground" naturalist and essentially a field biologist that he saw the danger of the rampant destruction and deforestation of tropical and forest life. In 1878, he published *Tropical Nature and Other Essays,* essentially warning those who would listen about the elimination of biodiversity for the sake of capitalism and industrialization.

Wallace, like Huxley, was certainly a social progressive. Perhaps because they were cut from the same cloth of poverty and family troubles they both wished to lend their support and advocacy to those less able to financially or socially support themselves. A self-described socialist, Wallace believed in the government leasing land to people, essentially providing access to land to address poverty and overcrowding in the cities. Wallace also aligned himself with the woman's suffrage movement and was a pacifist, claiming equal rights for all, and warning about the dangers and violence of warfare. He was well aware how modern war is so destructive to innocent civilians and society alike. Wallace was also worried about the Victorian penchant for crime and punishment, which did not center on inmate rehabilitation but the ongoing castigation of the incarcerated.

He continued to write throughout his life, and even had two books published in 1913, the year he died. He also lived long enough to enjoy the accolades that were due him. These included honorary doctorates from Oxford and the University of Dublin, several medals from the Royal Society, and medals from the Royal Geographical Society, the Linnean Society, and the Société de Geographie. He also had the great honor of receiving the Order of Merit from the Crown—an odd but nice turn of fate for an old radical.

After almost a century of life, Alfred Russel Wallace died at the age of ninety on November 7, 1913. He died at his country home and he was laid to rest in southern England in Broadstone Cemetery in the county of Dorset. Upon his death there was a movement to have him buried in Westminster Abbey alongside his colleague Charles Darwin, a place of honor based on his contributions to biology and science. But like Joseph Hooker, Wallace intended to be buried where he preferred and to be close to his family.

However, a white marble sculpture in the form of a low-relief profile of Wallace was placed in Westminster Abbey in honor of his contributions to science. Its location? Right next to the memorial for Charles Darwin.

CHAPTER 12

THE STORM: NINETEENTH CENTURY REACTION TO THE *ORIGIN OF SPECIES*

"Science adjusts its views based on what is observed. Faith is the denial of observation so that belief can be preserved."

Songwriter Tim Minchin

"There is no harmony between religion and science. When science was a child, religion sought to strangle it in the cradle. Now that science has attained its youth, and superstition is in its dotage, the trembling, palsied wreck says to the athlete: 'Let us be friends.' It reminds me of the bargain the cock wished to make with the horse: 'Let us agree not to step on each other's feet.'"

Robert G. Ingersoll

Introduction

The author Howard Zinn is frequently quoted as saying, "You can't be neutral on a moving train."[219] Certainly this phrase captures the essence of the reaction to *Origin* when it was published in 1859. No one who read or heard about the book was without an opinion about Darwin or his theory.

An international cadre of scientists and intellectuals, the British aristocracy, theologians from many faith traditions as well as lay persons weighed in and offered numerous opinions on Darwin's work. They reacted by writing critiques of the theory or whole articles and books saluting or refuting Darwin and his ideas. They debated *Origin* both publicly and privately at universities and learned societies. Discussion was not unknown in churches or in taverns. From school benches to the pulpit, and from the halls of power to the bar stool, the whole Western world would now have to come to terms in some way with Darwin and the theory of natural selection.

The *Origin* caused a significant uproar in scientific and religious communities, but it also had a major impact on the conservatives and the reformers within British society. Natural descent though random modification upended long held beliefs regarding the elite's innate, or in some cases divine view of their superiority over the masses.

The book also helped galvanize both Darwin's supporters and detractors. Those who read *Origin* correctly were either enthusiastic or horrified to think that natural selection was operating on all living species. They understood his theory, whether they agreed with Darwin or not, that we humans were not living apart from the forces in nature.

Then as today the view of natural selection, or descent through modification, falls into three separate if unequal categories of acceptance. There are those who outright deny evolution's operation on theological, biblical, or metaphysical grounds. There are people who accept the operation of natural selection as having been started and/or guided by a divine force (in most cases it is their personal god). Lastly, there are those who fully accept Darwin's theory as "truth" in that it explains materialistic nature through unbiased observation using methods firmly rooted in science. Until a better non-metaphysical theory can be vetted or operationalized it is Darwin's (and Wallace's) concepts that stand as the only plausible explanation for the development of all organic life on earth.

A Focus on the Notable and Knowledgeable

The list of people who wrote about and debated *Origin* when it was first published is amazingly diverse. One could make it their career and spend a lifetime tracking down every morsel of fact and data concerning the pivotal years of 1859 to 1875, when the theory of natural selection was in its infancy, and when it was still young enough to be politically and intellectually euthanized by those who saw Darwin and evolution as heresy.

There are dozens of museums and archives around the globe that maintain original and early documents, testimonials, and letters concerning Darwin and Wallace. There are many dozens of reputable websites with digitized correspondence concerning Darwin's friends and enemies alike. And many universities hold pieces of the Darwin puzzle within their hallowed archives and on the shelves of their numerous libraries.

Given constraints of both time and paper it is important to focus the chapter's discussion on the main proponents and opponents of Darwin. The reader is always encouraged to continue her or his research and use this book as a starting point or catalyst, rather than as a final destination regarding Darwin's life, his theories, and times.

The Theologians

To say that those in divine authority across Europe or outside its shores were in lock-step denial of Darwin's theory would be ignorant of the facts and historically inaccurate. The diversity of acceptance from clergy across Europe, America, and elsewhere was indeed varied. Of course, no clergy after reading *Origin* threw off their collars and priestly attires for a materialistic view of the world—at least nothing in the historical record has been found to show that this occurred.

But in more liberal theological circles the inclusion of natural selection only revealed to those believers the beauty of God's creation. This accomodationist view of evolution, which attempts to marry faith to science, is still very much alive today. However, those whose faith is more fundamentalist or evangelical are deeply threatened by evolution's claims.

Thus the strategy of "teaching the controversy," as modern day creationists and intelligent designers like to push, isn't a new strategy at all. It is one that harks back to the time before the Enlightenment when faith traditions ruled nations. Remnants of this political power, which once used to shape science investigation and scientific thought in the nineteenth century, are still alive today. But in the nineteenth century the divide between faith and biology immediately became hyper focused once biology tied itself to natural selection and the theory of evolution.

To the dismay of these conservative clergy, Darwin revealed that humanity owes more to nature, heredity and our biological ancestors than to anyone's personal sky-god, no matter how much they believe in such a deity. So it became the clergy's de facto responsibility to ally itself with scientists of faith and to then challenge Darwin's theory as quickly and as fervently as possible. But speed and intensity were just two points of the triangle. They also had to make these challenges as public as possible, all in an attempt to sway both scientific and public opinion.

The first formal debate shots over natural selection were fired and it was Darwin and specifically his apostles and advocates who came out more energized than when they went into those meetings. These were important battles won in one of the longest running scientific/secular wars ever produced by humanity. This debate about the nature of nature would foreshadow later controversies and aid in the "culture wars" throughout the nineteenth, twentieth and twenty-first centuries. Think most recently of the work of Hitchens, Dawkins, and Dennett, as well as the 2014 Bill Nye-Ken Ham debate for the latter.

Not All Anglicans Doubted Darwin—The Case for "Essays and Reviews"[220]

One may think this ironic, but it's clear that faith doctrines evolve in Darwinian ways like all man-made philosophies and social constructs. It has been argued by many, including the late Christopher Hitchens, that secular culture has been pushing all faith traditions toward modernity in numerous ways ever since the Enlightenment. So it is not a wonder to have within every faith tradition a liberal, conservative, and evangelical sect. Each interprets the same scripture differently and each has its own clerical and lay observers.

Enter the 1860 work entitled *Essays and Reviews,* written mostly by liberal Anglican clergy, and which was meant to move the church away from elements of superstition. The work included a stern denial of miracles as being unfounded and an affront to the divine. It was held that intervening miracles implied that an all-knowing creator got something wrong and thus had to change His path. Such change in direction would mean that God was not all knowing and infallible.

One of the chief theological architects of *Essays and Reviews,* the Reverend Baden Powell, thought that Darwin had shown through his theory the providence for God's work. God's handiwork in nature existed to show divine good and God did not need miracles to make life happen or to change it, so long as the process of natural selection that the deity created was in operation. Like American apostle Asa Gray, who wished to build a bridge between faith and science, Reverend Powell had hoped to deescalate the conflict between God, nature, and natural selection.

As noted previously, in 1889, the book *Lux Mundi* or *The Light of the World* was published. The book, by Anglican clergy, attempted to re-work Anglican theology through the prism of Darwinian natural selection. It sought to harmonize faith and science.

Other Religious Responses to Darwin and Natural Selection

An American clergyman, the Princeton Theological seminarian B. B. Warfield, held that accepting evolution was important to know God's work. He believed that understanding the natural mechanics of nature did not need to lead to an atheist worldview. Warfield's view, like that of modern-day Dr. Francis Collins, the eminent scientist, director of the National Institutes of Health and leader in genome science, lands squarely on the idea that one can have a successful career as a scientist without giving up their faith. Such bifurcation of reality and faith seems improbable to square, but is easily made for the scientist who must choose to believe in the divine and science simultaneously.

However, theologian and fellow Princeton seminarian Charles Hodge held an opposing view from Warfield (and that of Francis Collins). Hodge saw Darwin as essentially throwing divine creation into the dustbin of history. Accepting evolution means abandoning religious philosophy, thus leaving atheism as the only alternative narrative. Hodge felt faith and natural selection are mutually exclusive. Acceptance of the two cannot co-exist within one's personal philosophy or career as a scientist.[221]

More staunch American religious evangelicals rebuked Darwin's ideas and even those of Lyell in favor of what would become "Young Earth Creationism" or YEC. A catalyst for this idea was Ellen G. White, an evangelical who rejected Darwin and told all who would listen that she had visions from God and had personally seen the biblical creation. She also claimed that fossils could be easily dismissed as nothing more than remnants from Noah's Flood. As fringe-like and radical as they may seem, the YECs have grown in number and in some cases hold growing political power. Without them we certainly would not have the Creation Museum or the new Ark Park in the hills of Kentucky would have gone out of business some time ago.

Influential American pastor Enoch Fitch Burr, an astronomer by training who studied at Yale Theological Seminary, ardently rejected Darwin. Burr, a frequently outspoken critic of Darwin, saw the danger in materialistic evolution. Such challenge to theology only made Burr castigate Darwin's theory as being unrealistic for lacking at its core a supernatural creator. Burr, like many lettered believers in his day, refused to look at what science was telling them about the material universe, and he and others favored divine cause for the creation of all things.[222]

Anglican minister Charles Kingsley was an early proponent of natural selection. He quickly assumed the processes were exactly connected to the grace of God and his metaphysical ability to create and control life in the universe. For Kingsley, it was Darwin doing humanity a great service by allowing us to glimpse God's work. Kingsley was very active in the *Linnean Society* and throughout his life attempted to fuse his rather liberal interpretation of scripture with the science of the times.[223]

Nuanced responses from the clergy of the time included that of popular minister Charles Spurgeon. Minister Spurgeon could be at times openly hostile to Darwin in his sermons, particularly in his famous *The Gorilla and the Land He Inhabits*.[224] At other times he would use and make inferences to natural selection in sermons on social issues impacting the British people. The oratorically gifted Spurgeon used a Rorschach Test approach to natural selection, taking from it what he could in the moment to bolster his persona amongst the mass of his followers.

The Reverend Frederick Temple was actually in attendance at the Huxley-Wilberforce debate of 1860. A man of deep intuition, during Temple's lifetime he preached many sermons which urged the acceptance of natural selection. He concluded that evolution was not antagonistic to faith, choosing to see the truth in Darwin's theory and that of science, all for the sake of strengthening one's faith in the divine. Temple would go on to become the Archbishop of Canterbury.

The influential British missionary, David Livingstone, was not intense in his reproach to Darwin, but all the same did not see natural selection working on living species in Africa. Livingstone, of the famous "Dr. Livingstone I presume," certainly maintained his faith throughout his relatively short life. As a component of that faith came the denial, for purely theistic rather than material reasons, of natural selection.

From organized Catholicism, the first direct opposition to Darwin did not immediately come from Rome. However, a group of German Catholic bishops took it upon themselves in 1860 to write an open letter denying natural selection. The Provincial Council of Cologne's main point essentially stated that descent though modification violated the sacred faith and replaced God with nature. This pronouncement allowed the church to passively endorse an anti-Darwinian view, a view that could damn the believer to hell, all the while as it formally vetted *Origin* inside the halls of the Vatican.[225]

Prior to the bishops' statement, it is important to remember that there were many scientists who were also members of different Catholic sects who were instrumental in developing the early theories that would eventually lead Charles Darwin to develop natural selection. For instance, Jesuit Jean-Baptiste Lamarck developed the concept of transmutation and the Augustinian Monk, Gregor Mendel, essentially founded the science of modern genetics. However, post-*Origin,* the frequently read Jesuit periodical, *La Civilta Cattolica,* was decidedly anti-natural selection.

Out of lock-step coordination with the Vatican, former Anglican Priest turned Catholic cardinal John Henry Newman attempted to combine his view on theology with that of Darwin's theory. For Newman, evolution and Catholic doctrine were not inconsistent positions. Of course other impressionable Catholic intellectuals were less open to Darwin. The Catholic German writer Joseph Scheeben stated that evolutionary theory was a form of heresy if it were applied to humans.[226]

Indeed, Pope Pius IX, a strict conservative and defender of holy orders, pressed the First Vatican Council into accepting the concept of papal infallibility. This meant that an edict from any future pope was synonymous with the Church's dogmatic interpretation of doctrinal truth, a wholly authoritarian view held by many even today. The First Vatican Council stated specifically that all faithful Christians must reject science if such ideas and theories contradicted the doctrine of faith.[227] The Council was also clear that if a Catholic accepted anything other than God as creator then that person could and should be excommunicated.

Finally, it is important to briefly note other religious scientists and men of the cloth who accepted Darwinian evolution, in whole or in part, so long as it stayed within the primary bounds of a theological construct which placed the divine ahead of science. The accomodationists (or bridge builders if you prefer) included American geology professor Joseph Le Conte, New York theologians Lyman Abbott and James Iverach, the American clergyman Judson Savage and Protestant minister James McCosh, as well as pastor George Wright, a colleague of Darwin's apostle Asa Gray.

The Intellectuals React: Artists, Writers, Economists, and Philosophers

Just as Shakespeare would change our language with his numerous plays and poetry, Darwin's work would influence the science and non-science world around us as well. From literature to philosophy, from social thought to economics, the use of Darwin's ideas has made him a "prophet" or "boogieman." Since words have meaning and word meanings create ideas, it is no wonder that Darwin's impact has been so long and so varied within so many cultural and social contexts.

Words and concepts like "Darwinism" or "survival of the fittest" or "social Darwinism" or "descent with modification" and even the word "evolution" are all stuck to Darwin whether or not he used them in any real context within his writings and letters. As language evolves and word meanings and phrases change, even direct quotes attributed to Darwin may not be accurate or even his at all. So is the case found by Cambridge University's Darwin Correspondence Project,

> It is not the strongest of the species that survives, nor the most intelligent that survives. It is the one most adaptable to change.

The quote above, often attributed to Darwin, was in fact written one hundred years later for a science textbook. It became "Darwin's" when a reader who initially shared the quote removed the original attribution and subsequently it took on a life of its own.

The concept of "Social Darwinism" not only helped shape the later Industrial Revolution, but it also reinforced the idea of a class-based society, and set rich and poor against one another. Social policies based on this were all based on an incorrect interpretation of natural selection. Essentially, social Darwinism is an attempt to apply biological concepts of "favorability" and "survival of the fittest" to individuals and groups within society and economic systems. The result of such faulty understanding of Darwin's theory has led to many instances of racism, ethnic violence, eugenics, and imperialism over the last two centuries.

Karl Marx and Friedrich Engels, the social critics and economists, were both impacted by Darwin. It is said that Engels received his copy of *Origin* soon after it was published and read it with much interest. Materialistic naturalism is a core feature of evolution and socialist ideology, so it is no surprise that in this way Darwin, Engels, and Marx were simpatico at least in a macro sense. Physical reality stands on its own merits.

Marx's assessment of Darwin was certainly one of agreement, and in 1861 he wrote the following to a colleague:

> Darwin's work is most important and suits my purpose in that it provides a basis in natural science for the historical class struggle...Despite its shortcomings, [Author's note, Marx is referring to Darwin's writing style not his content] it is here that, for the first time, 'teleology' in natural science is not only dealt a mortal blow but its rational meaning is empirically explained[228]

However important Darwin was to both Marx and Engels, it was clear that inferring natural processes as somehow being allied to socialist ideology was never an exact fit, much in the same way that natural selection was incorrectly used by the Social Darwinists to enhance the upper classes and a particular race along with an imperialist ideology. Neither the political right nor left can make full use of *Origin* or infer some special alliance. This is simply because human social structure no matter what it is called or how it is defined and controlled is consciously managed by humans and not subject to unbiased nature.

Ironically, it was British sociologist Herbert Spencer who actually coined the phrase "survival of the fittest" and not Charles Darwin. Spencer metaphorically saw society as a living organism and thus considered it an entity subject to natural laws.

Spencer was a sociologist for his time and his theories sat well with the rich and politically connected. But his ideas did nothing to help those steeped in poverty. Because Spencer and Darwin's work appear at the outset to be compatible, in the nineteenth century, each had their theories used to justify and exploit social stratification—although Darwin never intended his work to be used in this way.

Friedrich Nietzsche, the existentialist philosopher and social critic, read *Origin* after it was published and his critique of Darwin's theory was pretty negative. While Nietzsche himself challenged Christian thought and values, he felt that natural selection left too much to randomness and the unconscious natural world. He also was not sure that the selection process worked on larger animals or humans. Nietzsche incorrectly thought that natural selection favored a "will to self-preservation" when actually all that happens is that organisms that don't die before they reproduce are represented in the next generation of that organism. Nietzsche, however, didn't like the "will to self-preservation," preferring for philosophical and aesthetic reasons a "will to power."[229]

It was Nietzsche who proclaimed "God is dead." But we know it was not old age or illness that ended the life of the divine. It was Charles Darwin's natural selection that killed him. We should remember that Nietzsche was not a scientist, but a social philosopher, so his criticism of Darwin was not based on scientific evidence. His challenges were essentially based on his biased thoughts and opinions. They may have had weight given they're from one of the greatest modern philosophers, but that does not mean Nietzsche was correct. Certainly both micro and macro-evolution has been fully vetted and evidenced in the fossil and biological record.

Many other intellectuals became early proponents of Darwin and natural selection. Social scientists and activists like the sociologist John Stuart Mill as well as the literary and theater critic George Henry Lewes, Victorian authors George Eliot, Victoria Gaskell, Thomas Hardy, and Charles Dickens each affirmed Darwin in their work and in some cases within personal letters to friends and colleagues. Certainly, Dickens's work shows the outcome of social Darwinist thinking and its impact on the poor and impoverished prior to social reform.

The late author Thomas Hardy's story is so reflective of our times. As a young man Hardy had religious faith but he lost it as he grew up and watched the church stand by and do little to help those without any means. In one of his later letters he writes, "My pages show harmony with Charles Darwin, T. H. Huxley, Herbert Spencer, David Hume and John Stuart Mill."[230] Nearing the end of his life, he wrote a poem entitled *Christmas: 1924*, about the inadequacies of faith and the church, and the folly of war.

In the United States, Darwin also caught the imagination of the literary intelligentsia as well. His writings, including *Origin* and his next book, the *Descent of Man*, had an influence on numerous American authors. These included William James, John Burroughs, Edith Wharton, and Herman Melville. Melville was a decade younger and deeply respectful of Darwin. Both men travelled to the Galapagos when young, and the islands left a lasting impression on both men.

The social scientist Lewis Henry Morgan, one of the fathers of American anthropology and his contemporary Benjamin Walsh, a former British citizen who immigrated to the United States, were also deeply impacted by Darwin's numerous writings, especially that of *Origin*. Walsh may best be remembered as a geographer while Morgan was famous for his almost Marxist theories on society. In 1877 Morgan illuminated his ideas by writing a treatise that broke societies into three categories of evolutionary development: savagery, barbarism, and civilization. As one can imagine, white European culture was clearly the "civilized" culture, lending credence to the faulty "survival of the fittest" mantra of economic and political social evolutionists of the nineteenth century.[231]

Finally, we should remember the advocacy of Francis Ellingwood Abbot, a Unitarian Universalist from Boston, who attempted to enroll Darwin in the *Free Religious Association*. Abbott saw natural selection as offering the opportunity to refute superstition, and since he served as one of America's notable freethinkers of his day, Abbot was an early ally to Darwin and *Origin*'s fundamental scientific theory. While Abbot may be remembered by many, he did have a unifying role in connecting many late nineteenth century freethought groups together in the United States. So his advocacy was, in part, very important in fomenting agreement that freethought and evolution needed to be connected to one another since each looked to the natural world to prove our non-mystical creation.[232]

British and International Scientists React to *Origin*

Much was written in the press starting a few days prior to *Origin*'s publication and then the floodgates of reviews and opinions poured in from around the globe, both for and against Darwin's work.

Sir John Herschel, the British eminent astronomer, mathematician and botanist, was said to call Darwin's book, "the law of higgledy-piggledy" squarely on the grounds that it removed divine providence from human existence. In America, Harvard's Jeffries Wyman, an anatomist and the President of *American Association for the Advancement of Science*, both privately and professionally rejected natural selection on both philosophical and theistic grounds.[233]

With some sadness to Darwin, his earlier mentor and advocate John Stevens Henslow was tepid at best when it came to praising *Origin*. Perhaps it was Henslow's wish to remain above the fray and a healthy skeptic? But in barely acknowledging the work and its meaning, his lack of activism would be seen by Darwin's detractors as a de facto repudiation of natural selection.

Silence was also the case with British scientist Michael Faraday. Widely renowned for his work in electromagnetism, Faraday never mentioned Darwin's work although he was keenly aware of it and its implications. Perhaps it was his deeply held religious beliefs or his illnesses in later years which stopped Faraday from commenting. But a man of such gravitas and intelligence had to know that by saying nothing, his lack of communication would be seen as an act of minimizing Darwin and his work.

Another esteemed British scientist, William Thompson, better known as Lord Kelvin, may have accepted evolution apprehensively and perhaps can best be seen as a follower of theistic evolution. Kelvin's objection to Darwin was based on his research on the age of the earth. According to Kelvin, the earth was perhaps only tens of thousands of years old; therefore it would be impossible for natural selection to be true. According to Darwin's theory, the earth must be millions of years old for the accumulation of traits and the natural creation and distribution of species. Hence, the immediate conflict between Kelvin and Darwin.[234]

Darwin believed that he had become the whipping boy for other established theistic scientists. Both Adam Sedgwick and Trinity College Fellow and zoologist John Willis Clark would frequently belittle Darwin and support each other. Each held firmly their distaste for natural selection.

Less tangential to Darwin was the brilliant scientist James Clark Maxwell, who was a deeply religious Christian and physicist. He certainly believed in divine creation and based this view on his research on molecules, which he described as "manufactured articles." He concluded that they could not form through the natural processes described by Darwin. Maxwell writes,

> No theory of evolution can be formed to account for the similarity of molecules, for evolution necessarily implies continuous change, and the molecule is incapable of growth or decay, of generation or destruction…None of the processes of Nature, since the time when Nature began, have produced the slightest difference in the properties of any molecule. We are therefore unable to ascribe either the existence of the molecules or the identity of their properties to any of the causes which we call natural.[235]

In Darwin's plus column was William Carpenter, writing in the *National Review,* who did a very positive review of *Origin*, and stated clearly that a pure theological refutation of Darwin's work was tantamount to a vapid response not only to Darwin's work but also to science. Darwin enjoyed the article so much that in January 1860 he wrote a letter of thanks to Carpenter.[236]

Other notable British scientists who were favorable to Darwin's work included botanist Hewett Watson and anatomist and naturalist Robert Grant. While Watson may have admired Darwin from afar, lauding him and his theory, Grant actually had known Darwin personally. It was Grant who had taken the young Darwin on field trips to study invertebrates.

Fellow countryman William Branwhite Clarke had moved from England to preach in Australia. While "down under" Clarke, upon receiving his copy of *Origin,* at once strongly concluded that Darwin's theory was not only correct but that it need not cause a schism for those of faith. Clarke was also a geologist and he felt that Australia was one of the more perfect places to research and study to prove Darwin right.

Another scientist Christian, Charles Babington, known in his time as an astute botanist and archaeologist, was convinced of Darwin's correctness. Babington accepted natural selection almost immediately after *Origin* was published. In fact, Babington and Darwin had a long history of letter writing and as contemporaries each vied to name and share species they found on their separate travels.

Still other scientists from across the globe challenged Darwin's theory for the sake of their own views regarding nature. In Germany, the anatomist Albert von Kolliker rejected natural selection for his own

Theory of Heterogeneous Generation, a form of punctuated equilibrium, where species change dramatically and quickly over a short period of time, versus Darwin's long and slow theory of common descent over millennia.

Another German scientist, Heinrich Brunn, archaeologist and director of the Glyptothek Museum in Munich, was one of the first scholars to translate *Origin* into German. While he had some pointed concerns and included his own criticisms in the translation, he correctly saw Darwin's view of evolution as becoming widely accepted in both scientific and lay circles.

The German biologist and naturalist Earnst Haeckel was immediately supportive of Darwin's work. As an early proponent he took *Origin* forward into German scientific and intellectual circles. This counter narrative to von Kolliker and the anti-Darwinists in Germany was similar to the debates happening all over Europe. Haeckel is probably best remembered for creating the concept of *recapitulation,* an idea that held a fetus, no matter the species, will go through stages which appear to mimic their much earlier phylogenic ancestors before being fully formed.[237]

In France, the physiologist educator and early developer of anesthesia, Jean Pierre Flourens, discarded Darwin's theory wholly on philosophical grounds. Flourens wrongly believed that Darwin had anthropomorphized nature, giving it human qualities and attributes. Flourens's issue was that he could not conceive of nature being unconscious in the way Darwin described how change over time occurs.

Conversely, the French positivist philosopher Emile Littre was a supporter of Darwin and saw humanity's place in nature as both real and valuable. Although not a scientist per se, he was a lexicographer. Littre saw deep meaning in the modernity that Darwin brought forward in *Origin.* He was a champion for human knowledge and viewed *Origin* as breaking down barriers in our perceived self-importance. A wonderfully famous cartoon of the time shows Littre holding up signs that Darwin crashes through in a circus-like setting, the idea being that Darwin's theories broke down human gullibility, superstition, and ignorance.

Still other scientists would reject Darwin's theory for reasons other than faith. Enter British zoologist George Mivart, a man who shared collegial relations with both Darwin and Huxley and who was at first a leading proponent of natural selection when Darwin first published *Origin.* But as time passed, Mivart grew weary of the theory from a

The famous Littre-Darwin Monkey Cartoon

purely design perspective, saying "clearly half of a wing serves no purpose," a critique not unlike those claimed from the modern intelligent design school of irreducible complexity pushed by Michael Behe. Mivart also wrote scathing critiques of Darwin's later works and in ways bastardized them by removing words from Darwin's sentences to change the meaning of his writings. Darwin rightly saw the attack as being unfair, hostile, poorly written, and without merit. Both Darwin and Huxley would never speak or write to Mivart again.[238]

There were a wide number of czarist Russian scientists who either accepted or denied Darwin's work. For the most part, those critical of Darwin almost uniformly came to the conclusion that his book was at best a working theory and that concepts like the "struggle of existence," his reliance on Malthusian concepts of population growth, and the view that all-too complex living conditions needed much more review and analysis to suggest that Darwin was right. However, those Russian scientists who accepted Darwin's theory, much like scientists who favored his other later works, were seen at outcasts and radicals at the time.

As noted in chapter 8, geologist Louis Agassiz, who should be recognized for his important work on ice ages, was another theologically inspired detractor of Darwin. Agassiz was born in Switzerland but settled in America to teach at both Cornell and Harvard Universities. Agassiz was deeply religious and while he disagreed with slavery, he did believe that *The Book of Genesis* proved only that white people were created by God. All other races, he claimed, came afterwards. Agassiz was affronted by natural selection as he felt the theory materialistically disemboweled the need for the divine, theological intervention or racial order.

Canadian geologist John William Dawson, a purist in thinking his whole career, found Darwin's theory troubling not so much for biblical reasons but because of the methods Darwin used to come to his conclusions. The fundamental element missing here was not the process of evolution, which Darwin articulated very well, but the underlying mechanics of evolution through genetic modification, something which Darwin did not have at the time of the publication of *Origin*.

This last criticism is especially stinging. Not because Darwin, who spent twenty years researching what would become *Origin*, was negligent but because the science of genetics was itself in its infancy. Gene theory just was not well developed or well known outside of those who read Gregor Mendel's work. The fact that Darwin and Wallace figured out the process of evolutionary biology without the fine details of gene mutation, gene drift, or gene flow was and remains amazing.

The Reformers—Use of Natural Selection to Change Social Ideas and Policy

As we will see in more detail in Chapter 14, the nineteenth century was an age of advances, hardships, and reforms. Technology was changing how we communicated and it was also helping to modernize as well as create whole new industries. Larger and larger factories, farms, and mines were being designed. Assembly lines with an untethered workforce increased both efficiencies and output.

The first real migration of people from the countryside and farms to the cities was growing daily. Cities were becoming increasingly crowded and crime-ridden. Illness was rampant and working-class families fell on hard times. Children were left on doorsteps of churches and private homes, whole communities of foundlings were ignored and received little if any formal education.

In England, the monarchy, the church, the political class and the owners of industries bolstered their wealth. At the same time, those who worked in the factories were treated poorly. Their mistreatment and misery was caused by the simple rule of supply and demand. The abundance of people willing to work cheaply far outnumbered the jobs available in any given industry. It was a bosses' buyer's market and the labor each person was selling was devalued. Without worker protections, if any individual left their job they could easily be replaced by the next person, or next hundred people, waiting in line for their position.

In fact, some activists who misread Darwin actually thought that he was advocating the use of natural selection to enforce social segregation. That simply was not the case. But the use of Darwin's theories, whether they were fully understood or not, served to enforce the materialistic ideas which motivated many activists to seek social policy change from the government or, in the words of Dr. Martin Luther King, Jr., a hundred years after Darwin wrote *Origin,* "be the change" themselves.

Many of the British reformers and activists used Darwin's ideas to push support for helping those in need. Perhaps they understood what many humanist and atheist activists accept today as true: that the materialistic nature of natural selection explains our physical reality and how little time we have on the planet. How small and insignificant we are, except, of course in our most intimate relationships and when we do good on behalf of others. Indeed, the nascent humanist movement began to stir because of the activities of secular philosophers and activists alike who studied Darwin and who sought justice for the impoverished.

They witnessed unfair and inhumane work and social conditions that the formal church either ignored, or in some cases colluded with to keep the working class in poverty. They saw a government and monarchy keep its distance for the sake of the status quo. They saw industry embrace *laissez-faire* capitalism and industrialization that almost exclusively benefited the owners. This same economic system also kept workers in dire straits with little choice but to maintain the "devil's bargain" to keep their fragile jobs to support their fragile families.

So who were these agents of change? The first British activist who should draw our attention is George Holyoake. Holyoake was a newspaper editor and, in 1851 some eight years before *Origin* was published, invented the word "secularism." He is also credited for adding the word "jingoism" to our dictionaries. Holyoake holds the distinction of being one of the last persons in Great Britain to be convicted of the "crime" of blasphemy. Holyoake accepted *Origin* as chapter and verse in defining the material world.[239] Then as now, Darwin's theory fits perfectly well within the secular view of life in which science rather than faith leads to truth. Holyoake is also credited with helping the working class by bolstering the cooperative movement in order to bring social and economic reforms to Britain. Both Holyoake and Darwin also had at least one very good friend in common, Harriet Martineau, a scholar of

Comte and early sociologist agitator in her own right.

Harriet Martineau was a firebrand. She had an amazing energy for research, writing and liberal politics and did all three very well. A Comte scholar, she also penned many articles concerning religion, women's and workers' rights. Martineau had a somewhat long history with the Darwin family, being romantically connected to Charles's brother Erasmus. When Charles returned from his voyage on the *HMS Beagle,* they spent much time together debating the issues of the day and comparing their notes, hers related to sociology and his on natural history. When Darwin published *Origin* some twenty years later, it was Martineau who prided herself with sharing his work with many in her intellectual circle, extolling its relevance and value to our modern world. Martineau also wrote about equal rights for the poor to vote—something that she was deeply criticized for by the various elite classes. One other thing she and Darwin had in common were lifelong, often debilitating illnesses, which frequently imposed on their ability to live out even fuller lives.

The eminent thinker, geologist, and climatologist John Tyndall was a very popular scientist who published more than 180 papers during his research career. Unlike many of the activists and agitators, because Tyndall was so charismatic, he was fondly accepted by the upper class and was in heavy demand for teaching and speaking positions his entire life. Tyndall was one of a small cadre of science activists who created the X-Club, of which some of the founding members included the apostles T. H. Huxley and J. D. Hooker. The X-Club sought to push the bounds of materialistic science forward and away from theology. In 1874, he presented his "Belfast Address"[240] in which he supported Darwin and forcefully stated that science should be removed from theology and that any alternative to scientific knowledge, especially that of religious doctrine, be considered immaterial to true knowledge.

The Church of England's Frederick Denison Maurice was a theologian of a new stripe and order for the nineteenth century. Not only did he not follow the tenets of conservative interpretation of scripture, but he in fact is known today as an early and main contributor to the Christian Socialist movement. This is a social justice and religious movement who's "Christ the journeyman and activist worker who lived and died to overturn the establishment" can be seen in the literature and activism of Jesuit priests such as Father Dan Berrigan and the slain Father Carlos Romero.

Maurice was dismissed from Kings College because of his views. While it is not clear he held an opinion on Darwin or natural selection, his liberal views and activism certainly supported those outside the church as they worked to bring social change to British society. In doing so, he helped force the hand of the elites (religious or not) to improve the working conditions of the poor. This allied effort helped the materialists who did read and knew Darwin's work and who supported his writings.

The world famous Florence Nightingale, the mother of modern nursing, was also a social reformer and activist whose work with the ill, the poor and soldiers on the battlefield would make her a pillar of modern medical care. One thing Nightingale and Darwin had in common were their varied illnesses, some psychosomatic, others real, but all detrimental to their personal lives and in some ways limiting their noteworthy careers.

While Nightingale and Darwin never met, she did read his work voraciously. Her interpretation of Darwin's work helped to foster her activism on behalf of the poor. She was a deeply religious woman with a liberal bent in interpreting both scripture and science. In reading Darwin's work, she essentially concluded that if we are a lone species left to the unconscious whim of a materialistic universe, without direct divine support coming to help us in the present or future, then we had best help each other and ourselves. To Nightingale, such inference would lead to saving countless lives and would certainly help define her as having led a good and moral life.

The Aristocracy and the Prime Minister

Finally, let us briefly turn our attention towards the British political elite. Queen Victoria's rule was marked by a deep interest in the occult and mysticism. As the head of the Church of England, the royal family also maintained a vested interest in continuing positive relations with Protestant clergy, as well as with clergy from other Christian faith traditions.

Queen Victoria herself would comment of the physical resemblance of apes and humans, something that clearly shattered her perceptions but fit well into the common ancestor/descent with modification points made by Darwin in his grand theory. Upon visiting the London Zoo years before the publication of *Origin*, it is documented that Victoria saw in orangutans our common humanity. She's quoted as saying, "The orangutan is too wonderful. He is frightfully, and painfully and disagreeably human."[241] Such insight must have stayed with Victoria her whole life as it would be impossible to ignore the external similarities humans and great apes share and not understand how the underlying complexities of life have led to such a diversity of species.

In 1859, shortly after the publication of *Origin,* the long serving MP and twice prime minister of England, Henry John Temple Palmerston, floated the idea that because Darwin's work was so important, he should be honored with a knighthood from the queen. It is unclear how receptive Queen Victoria was to Temple's proposition but, in the end, Bishop Wilberforce quashed the idea. Wilberforce was at times a one-man show, working diligently to discredit Darwin and his theory in the halls of power, in scientific circles and within the monarchy.[242]

Victoria's husband, Prince Albert, supported the idea and had advocated for Darwin's inclusion in this honor. Sadly, Albert would die less than two years after the publication of *Origin.* Because he and Victoria were so devoted to one another, perhaps a second attempt to offer Darwin a knighthood would have been possible and thus, successful. But in his lifetime, Albert stood as a counter-balance to the conservative nature of the church and that of *laissez faire* industrialization. He was a reformer and educator and a man with a keen interest in science, technology and modernity.

Other royals were less enthusiastic about Darwin and his ideas. This was mainly because Darwin's theories laid to waste any presupposed or divine conclusion that those on top of the social food chain, that is, the monarchy, the church and businessmen, held their place because God wanted or needed them to be there.

That they found themselves at high stations in British society meant that they needed to cling to one another for support. This self-fulfilling circle had to be propped up not only by birthright, but also by social policies and a culture that kept them in place. With worker and citizen revolutions occurring across Europe, with unrest in the colonies and British imperialism beginning to wane, it is no wonder members of the extended monarchy felt threatened by Darwin and *Origin.*

The Lady Aylesbury was taken back when told by a supporter of Darwin's theory that "your Ladyship and myself sprang from the same toadstool."[243] Left speechless, she would never again criticize Darwin, at least not in public. This same effrontery to special or divine creation still gets creationists, evangelicals, and clergy worked up in our own time.

Conclusions

While Darwin would never receive a knighthood he did become very publicly visible as part of the British zeitgeist. His face has adorned the ten-pound note since November, 2000, making his portrait synonymous with British history and identity. Down House is registered as a cultural landmark in Great Britain and a UNESCO World Heritage Site.

In 2008, the Church of England published a full apology to Charles Darwin, but like most Church mea culpas, they usually come hundreds of years too late. The Vatican only formally acknowledged Galileo's contribution to science, admitting he was correct, during the reign of Pope John Paul II. This was eight hundred years after his imprisonment for espousing scientific ideas that were originally viewed as treason and heresy by the Church. Here though, is the English Church's apology to Darwin,

> Charles Darwin: 200 years from your birth, the Church of England owes you an apology for misunderstanding you and, and by getting our first reaction wrong, encouraging others to misunderstand you still.[244]

Charles Darwin is interred at Westminster Abbey, a fitting final resting place of deep respect for the man so reviled by the church in his day. He is buried next to John Herschel and near Charles Lyell. His place of honor at the abbey is certainly solidified by other notable scientists who share the space as well. They include Sir Isaac Newton, Lord Kelvin, and most recently, Stephen Hawking, as well as other important men and women of letters and accomplishment. These include Geoffrey Chaucer, Margaret Cavendish, George Handel, Charles Dickens, Alfred Tennyson, Thomas Hardy, and Rudyard Kipling.

The reaction to *Origin* was not in any way monolithic. Clergy were affected and those who were conservative did all they could to belittle natural selection and coerce others to their cause. Still other men and women of faith accepted *Origin* and saw the natural mechanics of materialistic nature as further evidence of God's supreme grace. There were few if any middling opinions of Darwin's paradigm shifting work. Scientists immediately accepted Darwin's theory, either with caution or zest. For those upset or confronted by *Origin,* the denial came from a place of religious criticism or because it was difficult to divest from poorer, sometimes long-held competing scientific theories which natural selection had upended and superseded.

Intellectuals and reformers across the globe also reacted to *Origin* when it was published in 1859 and in the immediate decades that followed. They saw in Darwin's ideas the potential to powerfully alter social and economic policy. There was never any doubt among those who ascribed to a secular worldview that Darwin and his theory of natural selection were allies philosophically.

But from the perch of history, a place we all sit upon in the twenty-first century, it is clear that the most important and persuasive group most swayed by Darwin's work are those who inhabit the world of science. Those who work in biology, chemistry, geology, physics and astronomy overwhelming accept natural selection as the best and most elegant scientific theory put forward to understand the biological world.

Scientists accept natural selection not because they work in lockstep coordination, but because science is the search for the truth. Science is *about* challenging authority and dogma, especially scientific authority and dogma. To date, even those few remaining scientists who are skeptical about Darwin's work cannot support any other theory or show any compelling scientific evidence to discredit Darwin and natural selection. If such evidence were to be made available it would be easy to move on from Darwin.

Evolution remains the only theory of biological origins that produces testable predictions. These predictions have been tested repeatedly, and have given us ever-greater amounts of evidence for evolution. Darwin and Wallace have given us the insightful gift of understanding the operation of the biological world. The modern sciences of biology, ecology, physiology, medicine, chemistry, and so many other scientific areas of study have enhanced and been enhanced by our understanding of how evolution operates.

Science, especially those sciences founded on evolutionary biology, does not work on a wing and a prayer but through a test tube and a microscope. Darwin observed this as did Wallace, and as do all the scientists today who are committed to truthful observational and scientific investigation. And that is how we come to the next chapter of the book. Those apostles who answered the call in 1859 through their advocacy made sure that Darwin's work was not lost to the ages.

CHAPTER 13

THE CALM: THE APOSTLES REACH OUT, REACT, AND ADVOCATE — A CHRONOLOGY OF MAJOR ACTS AND SUPPORT OF CHARLES DARWIN

"*The improver of natural knowledge absolutely refuses to acknowledge authority, as such. For him, skepticism is the highest duties; blind faith the unpardonable sin.*"

T. H. Huxley

"*I expect to think that I would rather be author of your book* [The Origin of Species] *than of any other on natural history*"

J. D. Hooker (in a letter to Charles Darwin)

"*Natural selection is not the wind which propels the vessel, but the rudder which, by friction, now on this side and now on that, shapes the course.*"

Asa Gray

"*How can the Church be received as a trustworthy guide to the invisible, which falls into so many errors in the visible?*"

J. W. Draper

"*In less than eight years 'The Origin of Species' has produced conviction in the minds of a majority of the most eminent living men of science… its principles are illustrated by the progress and conclusions of every well established branch of human knowledge.*"

A. R. Wallace

Greatness is not usually born overnight or in isolation, as it nearly always needs to be nurtured. Intelligence, bolstered through experience, comes into its full bloom when we are emotionally cared for and physically protected. It also comes into its own and is much more efficiently realized as we are encouraged by our families, community and peers to achieve academic brilliance and other success.

Advocacy and support came to Darwin at many different points, throughout his entire life. We can certainly pinpoint the start of this nurturing with Darwin's discouraged yet ultimately supportive father, Robert Darwin. The elder Darwin saw his son's academic experiences as meager and meandering. He viewed Charles as an aimless wanderer, failing out of medical school, obtaining his divinity degree but still unfocused.

Still, as a well-to-do father, Robert reluctantly funded Darwin's travels on the *HMS Beagle*. His attention to the family's wealth also meant that the older Charles Darwin did not have to worry about money. The unconditional loyalty from his father would ensure Charles's own status as an English gentleman of means, a proud family man, and equally a man of science.

Darwin's professors and his favorite authors took an early liking to Charles and who saw in him a budding naturalist. As a young student, no one at the time could have ever guessed Darwin's ultimate and profound contributions to science. But these early mentors include men like John Stevens Henslow and Charles Lyell. Each had a huge impact through their teachings and writings on the young Darwin and their work would show much later in his own work.

Lyell's *Principles of Geology*, would be formational for Darwin. He read and reread the great geologist's ideas concerning stratigraphy, and the lengthy processes by which land underwent the physical accumulation of materials, as well as how the simultaneous withering of those land masses changed geography over time. It was Lyell's ideas and observations that showed Darwin that his own concepts regarding adaptation were plausible. Biological evolution as described by Darwin is a slow process and Lyell's theories on geology showed long sweeping periods of time, over millions of years.

It was John Stevens Henslow who convinced Charles's father to allow Darwin to take his fateful voyage on the *HMS Beagle* and to also finance his son's work while abroad. This early advocacy for Charles would have a huge impact on Darwin's later ability to collect specimens, synthesize his ideas, and allow the intellectual and physical space to grow as a naturalist and author. Darwin would later reject theories like transmutation, based on the specimens he collected and the two decades long research and observations he did prior to the publication of *Origin*. A particular debt of gratitude is owed to these men, firstly for igniting Darwin's imagination and secondly for encouraging his travels and research.

It should be acknowledged also that it was Sir Richard Owen who, after Darwin's travels on the *HMS Beagle,* supported the young naturalist by reviewing some of the specimens Darwin returned with from his voyage. However, Owen was not a supporter of Charles's later work and he'd always feel as if he'd been personally insulted by Charles because of the theory of natural selection. Since natural selection rejected Owens's ideas concerning transmutation, it would also remove the divine from both evolution and natural history.[245]

We can even see the impact of Darwin's grandfather, Erasmus Darwin, on Charles's psyche. Erasmus Darwin was a physician, poet, and naturalist. Erasmus died in 1802, just seven years before the birth of his grandson Charles. Even today Erasmus remains a deeply respected man, seen as someone well ahead of his time when it came to thinking about the natural world.

In addition, Erasmus was an abolitionist and although a "Lamarckian" in his views regarding species evolving towards some sort of perfection, he wrote some of the most important pre-evolutionary treatises before his grandson would tie all the pieces together about a hundred years later in *Origin*. In fact, as a point of Darwin family trivia, Charles would die eighty years and one day (April 19, 1882) after Erasmus's own death (April 18, 1802). The Darwin family was a true pillar of the English science establishment, a family affair and greatly in line with that of apostle Joseph Hooker's own family lineage and experiences.

Writers on natural history like Richard Owen, Erasmus Darwin, Jean Lamarck, and Lyell all had an impact on Darwin's eventual synthesis . However, it was Thomas Malthus, with his book *Essay on the Principle of Population,* which would help both Darwin and Wallace as they wrote about the natural world in a constant state of struggle for survival.

It was Malthus who noted that populations would forever increase faster than their food source, thus causing all species to survive based on their ability to maintain their access to resources.[246] Because of this idea, Darwin's own view of species variation, adaptation, and extinction could be played out on the field of every species' ecology combined with their ability to pass on favored traits to survive into the next generation, a full-on biological struggle for survival as never suggested before.

A Chronology of Support

In attempting to lay out the breadth and scope of the book as an accessible go-to reference for the Darwin reader, we propose this chapter have less of a narrative feel and so for the sake of time, it will offer a targeted chronology as to the allegiances and contributions of Darwin and the apostles to our world.

It should be noted that the correspondences and friendships were not just meant as academic or research support. The advocacy that Darwin counted on with his closest friends provided deep emotional support throughout his life.

These nineteenth century letters and other correspondences show a social web connecting personal and world events, mundane issues, awards for and advocated by Darwin and catty discussions of well-liked and disliked obstructionist scientists of the time. They also include the discussion of rumors, pettiness, the sharing of joy that comes with personal success and also with empathy at loss. Literally thousands of letters between the apostles exist.

There is an interesting immediacy to these messages that makes reading them sometimes feel like reading email. But in the nineteenth century, each message would take days (sometimes weeks) to reach its intended reader. It should be noted with great thanks that these letters can be found readily online via the Darwin Correspondence Project, the Wallace Correspondence Project, and the Hooker Correspondence Project.

Offering a chronology also offers the reader insight into the clear tenacity of the apostles' support and their willingness to help Darwin throughout his years prior to and after the research, revision, and publication of *On the Origin of Species,* as well as Darwin's other later books.

Darwin and His Apostles' Lives Prior to the Publication of *The Origin*

1828—**The young Darwin (as a student) begins a letter writing to both Sarah and Fanny Owen** concerning social events that he's attended or missed with them and while at school. He also begins writing to his family members regarding his collections of bugs and other animals that he wishes to study and classify.

1828/29—**Charles writes to his elder brother Erasmus Darwin** (so named for his grandfather) about his studies, and his time at school and deepening friendship with roommate William Darwin Fox. Fox was Charles and Erasmus' second cousin. William and Charles shared a devoted interest in etymology and natural history.

1829—**Darwin writes to Fox about a trip to retrieve Entomolog-**

ical specimens gone bad, perhaps a prelude and first glimpse of his real or imagined health issues once he's stressed. Darwin writes, *"My dear Fox...I should have written to you before, only that whilst our expedition lasted I was too much engaged...I started from this place about a fortnight ago to take an Entomological trip....the two first days I went on pretty well, taking several good insects, but for the rest of that week, my lips became suddenly so bad, & I myself not very well, that I was unable to leave the room, & on the Monday I retreated with grief & sorrow back again to Shrewsbury."*

1830—The young Darwin writes to his cousin William Darwin Fox concerning both entomology and the need to have better storage for his findings. He is excited to know that a cabinet he's ordered for his growing specimens will be delivered shortly. (*Author's note: What seems like trivial correspondence portends a lifetime of observation, collection, description, and detailed analysis of a wide range of species accumulated by Darwin over his lifetime.*)

1831—Darwin writes to his sister, Caroline about his studies and about his growing relationship with John Henslow. By this time Henslow is a growing influence and mentor to the young Darwin. Darwin writes, *"...my enthusiasm is so great that I cannot hardly sit still on my chair. Henslow & other Dons give us great credit for our plan: Henslow promises to cram me in geology.—I never will be easy till I see the peak of Teneriffe and the great Dragon tree."*

1831 (August)—Darwin receives a letter of greetings and invitation by George Peacock to join the crew of the *HMS Beagle*. Darwin's role will be to serve as the ship's naturalist. Henslow of course has made this connection and supports Darwin's place on the ship.

1831 (September)—Darwin writes to Henslow concerning his meeting with *HMS Beagle's* **captain,** Robert FitzRoy, who has now officially offered the position of "Naturalist" for the coming voyage.

1831—The HMS Beagle sets sail from Davenport on December 10. Originally set to sail in late September under Captain FitzRoy, delays caused by re-fitting the ship's mechanical systems and deck (this was the ship's second voyage) leave the ship in port. However, upon leaving, the ship and crew will not return to England until October, 1836.

1832–1836—Darwin as ship's naturalist spends his days collecting and classifying the specimens he's accumulated during the whole five-year journey of the *HMS Beagle*. Darwin meticulously notes his specimens, sends letters to Henslow and to Hooker, as well as to his family and others about his finds. Darwin sends specimens to England.

He also writes about his health, the nature and conflict with Captain FitzRoy, and his fascination with his work.

1832—John Draper moves with his family to the United States. He settles in Virginia and will begin his medical studies shortly, before embarking on a career as a chemist, photographer, and physician.

1837—Ornithologist John Gould discovers that what we now call "Darwin's finches" were indeed all finches. Darwin was unaware of this fact and had assumed that all of the variety of bird specimens from the Galapagos were different bird species. This was indeed not the case, and it was Gould who investigated and shared his analysis with Darwin. This discovery was fundamental to the development of Darwin's ideas about evolution.

1837—Darwin receives a letter from his youngest sister, Emily Catherine, concerning the *Morning Herald's* article about Darwin's specimens now housed from his Galapagos journey. The paper recounts some 80 mammalian and 450 bird specimens now housed at the Zoological Society. This article may show the importance of Darwin's work as a naturalist and how he'd become a bit of a celebrity —although, as Emily writes, their father awaits the day when the Zoological Society will formally acknowledge Charles's contributions. Darwin's father, always a stickler for evidence and respect, clearly worries that Charles' work will go unrecognized by the scientific community of the day.

1838—While on his trip to England, Asa Gray meets with Darwin to discuss emerging theories related to ecology and natural selection in its broadest and still-defined terms. Although Darwin is decades away from publishing, his friendship with Gray will support his research over the next twenty years. Botanist Joseph Hooker, who is a casual friend to Darwin, the naturalist and Gray, a young American botanist, arranged this meeting.

1839—Darwin marries his first cousin, Emma Wedgewood

1839—Darwin's first monograph, *Journal of Researches* is published

1839—Darwin's correspondence with Gray widens as Charles continues to discuss North American specimens collected by Gray. Darwin and Gray would become collaborators on their individual works related to botany in the decades ahead.

1839—Joseph Hooker, while preparing for his journey to the Antarctic, is given a draft of *Voyage of the Beagle* by Charles Lyell for editorial review prior to its publication in late 1839.

1839—Joseph Hooker joins the *HMS Erebus* for what would become a four-year expedition to collect botanical samples in the Antarctic. The ship will travel extensively and return to England in 1843.

1839—Darwin publishes his first volume on his collecting and classification work while on the *HMS Beagle*. He would complete all five volumes of his expedition by 1844.

1840—Darwin continues his correspondence with Asa Gray concerning his North American botanical finds.

1842—Charles and Emma move to Down House in Kent, from their flat in London. Here the Darwin family will grow and the house and land will serve Darwin the rest of his life as a living laboratory. It is here that Darwin will observe and write his most famous books and articles, and share correspondence with the apostles and many others.

1843 (March)—Darwin writes to William Hooker to thank his son Joseph for his letter of interest in his work. Joseph writes to Darwin while he is on expedition in Antarctica.

1843—Charles Darwin begins to write to Joseph Hooker on many matters of science and nature. Their fast and deep friendship is solidified by their interest in science, botany, and natural history. Hooker becomes Darwin's "Librarian of Botanical Science."

1843—(December)—Hooker writes to Darwin thanking him for sending Galapagos plant species in such abundance for his review. Hooker writes in part, *"The Galapago plants are far more xtensive in number of species than I could have supposed, & are the foundation of an xcellent flora of that group: Mʳ Henslow as sent with them those of Macræ."*

1844 (January)—Darwin writes to Joseph Hooker and admits that he believes that *"species are not immutable,"* essentially writing for the first time to a colleague the first key idea of natural selection, that species change and adapt over time.

1845—Darwin writes to Joseph Hooker as a return correspondence to Hooker's earlier letter describing his career options and in part regarding natural history. Darwin writes, *"I am much obliged for your very agreeable letter; it was very good natured, in the midst of your scientific & theatrical dissipation, to think of writing so long a letter to me. I am astonished at your news & I must condole with you in your present view of the Professorship."*

1845—Darwin writes to Lyell about Lyell's new book, *Travels in North America,* and offers a personal friendly review. Especially interesting to Darwin is Lyell's dislike of American clergy. Darwin writes, *"Your account of the religious state of the States particularly interested me: I was surprised throughout at your very proper boldness against the clergy. In your university chapter, the clergy & not the state of Education* *are most severely & justly handled; and this I think is very bold, for I conceive you might crush a leaden-headed old Don, as a Don, with more safety, than touch the finger of that corporate animal, the Clergy."*

1846—Thomas Henry Huxley at age twenty begins his journey at sea as an assistant surgeon aboard the *HMS Rattlesnake*. The trip will take him from England to New Guinea and Australia; almost immediately upon arrival he begins sending specimens home. Although not the ship's naturalist, he continues to collect species and document the flora and fauna. Huxley and the ship return to England in 1850.

1847—Darwin writes to Hooker to meet one final time before Hooker departs for his Himalayan expeditions. Dinner is planned.

1847—Hooker begins what would be at times a harrowing expedition to the Himalayas. The trip, while fruitful and spent in headlong discovery, is also harrowing. At one point in the journey, Hooker and a mate will find themselves prisoner and may be at times close to execution in Tibet. Negotiations take place and Hooker and crewmate are eventually released unharmed.

1848 (February)—Hooker writes to Darwin about his arrival in India and of feelings of loss in terms of not being close to Charles. Hooker writes, *"Though our correspondence has not ebbed so low for full four years, you have been so constantly in my mind that it appears far from strange to be writing to you."*

1848—Alfred Russel Wallace embarks on his first scientific trip abroad to collect species samples in South America, first working in Brazil and then spending much of his time along the Amazon River collecting samples of flora and fauna. His trip will last for years and he would return to England in 1852.

1850 (January) Darwin writes to William Hooker thanking him for his update on Joseph's safety after being kidnapped while in Tibet. Darwin writes, *"I write merely to thank you very sincerely for your great kindness in taking the trouble to inform me about your son. I was very anxious to hear something about his safety."*

1851—Joseph Hooker returns from his journey to the Himalayas. He will bring back a trove of botanical samples from India and within four years' time begin to serve the Royal Botanical Gardens at Kew.

1852—Alfred Russel Wallace returns from the Amazon. During his return to England his ship, the *Helen*, will catch fire and have to be abandoned. The majority of his specimens will be lost to the fire. It is a blow to the early career but not the tenacious character of Wallace, who will continue his research and writing, eventually becoming known as the co-discoverer of natural selection.

1853—Darwin pens two letters to T. H. Huxley (both in April), at once praising him on his work with barnacles and mollusks and at the same time asking somewhat fitfully to read his own book on the matter. Their work and interests are similar as are their descriptions and thoughts concerning the natural history of barnacles. [*Author's note: Even here you can see Darwin's concern for a united front in the coming modernity wars.*]

1853—Darwin writes to Hooker about several issues concerning Hooker's recent work, Darwin also goes on record noting his disbelief in the permanence of species. While embedded in the larger letter, the line is significant as it shows Darwin's coming to terms with his ideas about evolution, adaptation, and extinction.

1854—Joseph Hooker publishes his book *Himalayan Journals,* and dedicates it "To Charles Darwin by his affectionate friend Joseph Dalton Hooker."

1854—Wallace leaves England for what will be an eight-year expedition across Indonesia. His travels, collections, and experiences will serve as the core narrative for his book *The Malay Archipelago.* While traveling he will write successful papers and communicate with many of the apostles, especially after his letters to Darwin show he has developed the same theory regarding natural selection based on his earlier research in the Amazon.

1854 (February)—Darwin writes to Hooker praising *Himalayan Journals* and thanking him for the book's dedication. Darwin writes, "*I do not know what to say—I have within these few minutes received your Book, & after admiring some of the illustrations, I fell on the Dedication to me. My dear Hooker, you have gratified me, more than I can speak, but really it is altogether too grate a compliment to me...*"

1855—Alfred Russel Wallace publishes a paper on his theory of species evolution based on his own research in South America.

1855—Wallace's "Sarawak Law" paper in the *Annals and Magazine of Natural History* focusing on the elementary ideas of natural selection (without using these words) arouses interests from many naturalists. Edward Blyth, a zoologist and friend of Darwin, writes to Charles noting, "*What think you of Wallace's paper in the Ann. M. N. H.?1 Good! Upon the whole!...Can we suppose a lost series of gradations connecting these genera with the Deer type, & ramifying off to them paulatim? Wallace has, I think, put the matter well; and according to his theory, the various domestic races of animals have been fairly developed* *into species.* Lyell also noticed this paper and brought it to Darwin's attention, pointing out that Wallace was close to developing the theory of evolution by natural selection, independently of Darwin.

1855—Hooker is appointed Assistant Director of the Royal Botanical Gardens at Kew. During this time, he will work under his father, William, who has served as Director for the institution for decades.

1856–1858: The First Strong Efforts to Support Darwin and his Grand Theory

1856—Darwin writes to T. H. Huxley, inviting him to Down House for an evening with Joseph Hooker and his wife, and other guests. This is a gathering of naturalists meant to spur the cross-pollination of ideas.

1856—While not an "apostle," eminent geologist and Darwin mentor Charles Lyell writes to Darwin, praising him for his work and urges him to publish his theory even without full evidence. Darwin the ever-cautious scientist continues to wait, observe, and rewrite.

1856—Darwin responds to Lyell's letter. Darwin writes, in part, the following:

> With respect to your suggestion of a sketch of my view; I hardly know what to think, but will reflect on it; but it goes against my prejudices. To give a fair sketch would be absolutely impossible, for every proposition requires such an array of facts. If I were to do anything it could only refer to the main agency of change, selection,—& perhaps point out a very few of the leading features which countenance such a view, & some few of the main difficulties. But I do not know what to think: I rather hate the idea of writing for priority, yet I certainly shd be vexed if any one were to publish my doctrines before me.—Anyhow I thank you heartily for your sympathy.

1856—Darwin writes to Hooker, asking him to review chapter three of Darwin's proposed book titled *Natural Selection*, asking for his comments and changes to the chapter. Darwin writes, *"Will you return this to me. with any remarks? ... Can you illuminate me? For this in my present state of ignorance seems the strongest case of 'Darwin, an eternal & necessary hermaphrodite.'"*

1857—Darwin pens a letter in response to Alfred Russel Wallace's letter concerning Wallace's work and findings, especially as they relate to the core concepts of natural selection. However, this is not the bombshell paper which will push Darwin into publishing *Origin*. Darwin writes, "*I am much obliged for your letter of Oct. 10th from Celebes received a few days ago: in a laborious undertaking sympathy is a valuable & real encouragement. By your letter & even still more by your paper in Annals, a year or more ago,3 I can plainly see that we have thought much alike & to a certain extent have come to similar conclusions. In regard to the Paper in Annals, I agree to the truth of almost every word of your paper; & I daresay that you will agree with me that it is very rare to find oneself agreeing pretty closely with any theoretical paper; for it is lamentable how each man draws his own different conclusions from the very same fact.*"

1857—T. H. Huxley offers three lectures over the course of late 1857, each in many ways praising Darwin's work in natural history and the classification of animals. In particular Huxley lauds Darwin's long research with Cirriipedia (barnacles). It is clear that Huxley is a huge supporter of Darwin and has been influenced by Darwin in his own research and writings concerning zoology and natural history.

1857—Darwin writes to T. H. Huxley thanking him for his supportive lectures. Darwin especially notes, "*I must just thank you for your three last Lectures which I have read with much interest (& have forwarded to J. Lubbock), & for your magnificent compliment to me.—*[2] *I declare you will turn my head right round. You have given, as it seems to me, a capital account of the Cirripedes*".

1858—Darwin receives a copy of Alfred Russel Wallace's now-famous essay from Ternate that proposes evolution by natural selection (though he does not use those exact words). Wallace's theories are duplicative to Darwin's almost twenty-year work on the theory of natural selection.

1858—Darwin writes to one of his mentors, Charles Lyell, about Wallace's draft paper concerning the operation of natural selection. He is deeply concerned that with publication, Wallace in small sketch will in essence beat Darwin's life work in the annals of scientific discovery. Darwin writes about this crisis:

> I sh[d] *not have sent off your letter without further reflexion, for I am at present quite upset, but write now to get subject for time out of mind. But I confess it* never did occur to me, as it ought, that Wallace could have made any use of your letter.[2]

> There is nothing in Wallace's sketch which is not written out much fuller in my sketch copied in 1844, & read by

Hooker some dozen years ago. About a year ago I sent a short sketch of which I have copy of my views (owing to correspondence on several points) to Asa Gray, so that I could most truly say & prove that I take nothing from Wallace. I sh^d *be extremely glad* **now** *to publish a sketch of my general views in about a dozen pages or so. But I cannot persuade myself that I can do so honourably. Wallace says nothing about publication, & I enclose his letter.—But as I had not intended to publish any* sketch,[5] *can I do so honourably because Wallace has sent me an outline of his doctrine?—*

I would far rather burn my whole book than that he or any man sh^d *think that I had behaved in a paltry spirit. Do you not think his having sent me this sketch ties my hands? I do not in least believe that that he originated his views from anything which I wrote to him.*

1858—Joseph Hooker and Charles Lyell arrange for the reading of Darwin and Wallace's twin and deeply similar theories on species evolution at the Linnean Society. Both Lyell and Hooker read the papers in support of Darwin, Wallace and natural selection.

1858—Both Darwin and Wallace write separate letters of appreciation to Hooker, who has remained a ready advocate of both naturalists and their conclusive theory. It is Wallace though who sees himself as junior to Darwin, and who is humbly grateful to receive the credit he does regarding natural selection. Wallace writes, "*I beg leave to acknowledge the receipt of your letter of July last, sent me by Mr. Darwin, & informing me of the steps you had taken with reference to a paper I had communicated to that gentleman.*[2] *Allow me in the first place sincerely to thank yourself & Sir Charles Lyell for your kind offices on this occasion, & to assure you of the gratification afforded me both by the course you have pursued, & the favourable opinions of my essay which you have so kindly expressed. I cannot but consider myself a favoured party in this matter, because it has hitherto been too much the practice in cases of this sort to impute all the merit to the first discoverer of a new fact or a new theory, & little or none to any other party who may, quite independently, have arrived at the same result a few years or a few hours later.'*

1859: That Fateful Year

For almost the entire year prior to publication, Darwin will send parts of the draft manuscript of *Origin* to several people, including all but one of the apostles for review, comment and editorial challenge and scientific guidance.

1859 (October)—Darwin writes to Hooker that he is sending four copies of *Origin*, prior to publication, to his dear friend, confidant and editor. One is for Hooker's shelf and three others are to be shared with "foreign botanists" as Hooker would employ in offering the work to colleagues.

1859—First Edition of *Origin* **is published** and sells out on its first printing. The book shakes up British and international science and faith communities and is debated in the British intellectual, political and lay circles.

1859—T. H. Huxley writes an anonymous and strongly favorable review of *Origin* in the Times.

1859—Huxley publishes *Time and Life: Mr. Darwin's "Origin of Species,"* in *Macmillan's Magazine.* The article is intended to more widely disseminate Darwin's contributions and ideas to a wider, and in some cases a non-scientific audience noting, *"In either case the question is one to be settled only by the painstaking, truth-loving investigation of skilled naturalists. It is the duty of the general public to await the result in patience; and, above all things, to discourage, as they would any other crimes, the attempt to enlist the prejudices of the ignorant, or the uncharitableness of the bigoted, on either side of the controversy."*

1859—Huxley publishes *The Darwinian Hypothesis,* an article supporting Darwin's work and noting that while there may be unknowns in his work which will be challenged by Darwin's detractors, that *Origin* serves as the best way to know the evolution of life and natural world.

1859 (December 1859)– Huxley publishes an anonymous and very favorable review of *Origin,* in the *Saturday Review,* noting the importance of the work. *"In regard to that which is peculiar to Mr. Darwin's theory, we are far from thinking that the fruits of his labour and research will be useless to natural science. On the contrary, we are persuaded that natural selection must henceforward be admitted as the chief mode by which the structure of organized beings is modified in the state of nature."*

1860 and Beyond—The Apostles (and others) Stand for Darwin as He Did Not Stand for Himself

1860—Darwin writes a letter of thanks to William Carpenter, a respected professor of physiology, for his favorable review of *Origin* in *The National Review.*

1860—Darwin will write several letters to Charles Lyell for his support and encouragement of *Origin,* even as they plan a second edition to include revisions and also the inclusion and prominence of Wallace's contributions.

1860 (March—May), Asa Gray vigorously defends *Origin* in numerous debates at The American Academy of Arts and Sciences. Gray essentially shows that a theist scientist can and should embrace natural selection, refuting Agassiz's furious belief that Darwin's work expressly supports an atheistic worldview. Agassiz also contended that Darwin's work was a fad, something Gray deeply rejected.

1860 (January)– Asa Gray writes to Darwin, warning him that Agassiz will be minimizing and publicly attacking *Origin*. In a letter dated January, 1860, Gray notes: *"and it may be useful, since it is largely occupied with a defence of you against Agassiz—who has been helping the circulation of your book by denouncing it as atheistical in a public lecture!3 I suspect, also, he means to attack it in the* Atlantic Monthly.*4 The book annoys him; and I suppose the contrast I run between his theories and yours will annoy him still more."*

1860 (May)—Darwin writes to Gray thanking him for his support of *Origin* since he is too ill to do anything but stand fast and let others debate his work. *"But the effect on me is that I will buckle on my armour & fight my best. You seem to have done so allready in grand style. And I believe Hooker will, as certainly will Lyell & Huxley. But it will be a long fight. By myself I sh^d. be powerless. I feel my weak health acutely, as I cannot work hard."*

1860 (July)—Asa Gray publishes his "lighthearted" book review of *Origin*, in *The Atlantic*. The review firmly gives Darwin's new book and theory justifiable acceptance in the American scientific, philosophical, and intellectual communities. The article also refutes misinterpretations and intentional misrepresentations of natural selection; it is lauded by Darwin and the other apostles in England.

1860—Huxley publishes more support for *Origin*, a more forceful and less controversial review of Darwin's work than the 1859 *The Darwinian Hypothesis*. The article firmly supports natural selection and its conclusions. It also supports Darwin as a leading scientist and naturalist of his day.

1860—The Huxley-Wilberforce debate takes place at the Royal Society . On the side of modernity and science you have John Draper, T. H. Huxley, and Joseph Hooker. On the side of religious-based biology and faith, the one and only Bishop Wilberforce, and the Captain of the *HMS Beagle*, Robert FitzRoy, among others. The debate moderator was the surprisingly fair John Henslow.

No full transcript of the debate is known to exist. Several letters to Darwin, descriptions by the apostles and others in attendance, and newspaper accounts note that the most striking debate points related more to insults rather than content (at least from the anti-Darwinists).

The most famous insult cast by Bishop Wilberforce came in his summation, in which he asked Huxley on which side his grandparents were apes. Ever-ready to play offense, Huxley retorted that he'd prefer to have an ape as a grandfather rather than be considered human if it meant being related to Wilberforce and his willful ignorance.

John Draper opened for modernity and while his talk was important and positive it was poorly received, and something of an undercard to the real match ahead. And while Huxley held his own against Bishop Wilberforce, it was actually Joseph Hooker who made the most salient and important points in favor of Darwin and natural selection.

Robert FitzRoy, aside from holding a Bible over his head and commanding everyone to believe in Christianity and accept the Christian God at face value, was not much help to anyone during the actual debate. His pro-religious stance was too dogmatic.

1860—Darwin himself writes to Asa Gray about Hooker's letter and the Huxley-Wilberforce debate. Here is in part the text of the letter:

> Yesterday I had letter from Hooker at B. Assocn. at Oxford;4 & he tells me that there was one day a savage fight on my Book between Owen & Huxley; & subsequently a discussion of utmost warmth of 4 hours duration(!) on a paper by Draper of U. States on some subject,5 in which somehow (I know not how) my book became subject: Bishop of Oxford, one of most eloquent men in England, ridiculed me at great length & with much spirit; & Hooker answered him, I imagine, with wonderful spirit & success.

1861—Wallace writes to Darwin regarding the ongoing controversies and successes of *Origin*. Wallace notes that while he hasn't been home to support the theory, he has advocated and given both copies of *Origin* to enlightened people in and around Indonesia where he has been collecting specimens for almost a decade. (*Editor's note: Because Wallace was away from England for the critical years 1859-61, after the immediate publication of* Origin, *he could do little to defend* *the theory. As such, as co-discoverer, he received little of the attribution or the criticism cause by Darwin's publication.*) Wallace left England in 1854 for Indonesia and would not return until late 1862.

1862 (May)—Darwin publishes *"Fertilisation of Orchids,"* in which he uses his ideas regarding natural selection in detail to enforce the ideas and concepts of descent through modification and speciation.

1862—Fellow naturalist and botanist Asa Gray writes a glowing review of *Orchids* for *The American Journal of Science.* The review praises Darwin's tenacity as a researcher and scientist and intensely supports natural selection.

1862 (November)- Hooker writes to Darwin and admits he penned two favorable reviews of *Orchids* which ran under the byline of another person who had become too unwell to write. The reviews appeared in *Gardeners' Chronicle* and *Natural History Review.*

1863—Huxley publishes *Man's Place in Nature,* using Darwin's ideas concerning natural selection to bolster his research. This book squarely places humanity's place within the context of animal evolution.

1863—Hooker writes to Emma Darwin concerning Charles's ill health and his overall decline in wellness. Emma had earlier written to Hooker of Darwin's failing health. He asks Emma if he should write directly to him or if receiving letters of kindness is even too difficult for Charles to handle.

1864—Huxley publishes *Criticisms on The Origin of Species,* essentially refuting the attempted discrediting of *Origin* by European scientists resistant to the idea of descent through modification. He specifically challenges the idea that Darwin's work discounts metaphysical intervention.

1864 (November)—Huxley creates the *X-Club,* **a meeting space for progressives, intellectuals** and the social and scientific elite who share ideas and support each other. Early first members include Joseph Hooker, T. H. Huxley, and Charles Darwin. X-Club membership was exclusive and elitist in many ways.

1864—Hooker writes to Darwin about the unfair exclusion of ***Origin*** in the awarding of the Copley Medal at the Royal Society. Both Hooker and Huxley in their comments at the meeting hope to ensure the relevance of Darwin's work in the scientific society. In a later letter (December, 1864), Darwin thanks both Hooker and Huxley for their continued support and is amused but unsurprised by his obvious exclusion.

1864—John William Draper publishes *History of the Intellectual Development of Europe,* an early version of which formed the basis of his remarks which preempted the Huxley-Wilberforce debate. Those remarks and the book favored Darwin's *Origin*. Infused in the talk was Draper's idea that Darwin's evolutionary view could be expressed in society as well. Draper expresses doubt that Darwin's work (and indeed our modernity) would be possible without the redirection of thought towards science and reason and away from religious faith.

1865—Darwin writes to Hooker and consoles him on the lack of acknowledgement for Hooker's father at the Royal Society. Darwin suggests that the elder Hooker's contributions to science are being overlooked because there were so few botanists within the Society's governance.

1865—Joseph Hooker succeeds his father, William, as Director of the Royal Gardens at Kew. His successful service is spent building the gardens even as his tenure is spent at times battling the allies of modern science and modernity who dislike him personally as much as Darwin's theory of natural selection.

1866 (August)—Hooker delivers a lecture on the *Darwinian Theory* to more than two thousand people in attendance at *British Association for the Advancement of Science* in Nottingham, England. The lecture is well received by all in attendance.

1867—The full text of Hooker's speech to the *British Association for the Advancement of Science* is published for wide distribution through *Gardeners' Chronicle* with the title of "Insular floras." It marks the continued advocacy, acceptance and strengthening of Darwin and Wallace's theory.

1867—Darwin sends pre-publication of *Variation* to Asa Gray for review and comment.

1868 (January)—Asa Gray writes back to Darwin explaining that he'd been busy but that he'd read up through four hundred pages of the *Variation* monograph. Gray's comments relate little to the content of the book but certainly this is not meant as criticism.

1868—Darwin publishes *The Variation of Plants and Animals Under Domestication*. The book's first edition--all 1,500 copies--is sold out within a week.

1868—Most of the apostles write to Darwin thanking him for his new book. Each also praises the work and promises Darwin that their next book to read is his. Huxley, always the most metaphorical of the group, claims that he needs a body with two heads or one that requires no sleep to keep up with his reading.

1869—Huxley publishes *The Genealogy of Animals,* a collection of essays devoted to the concept of phylogeny and how evolution of animals and plants showing clear descent from each other create a phylogenic family tree.

1869—Wallace publishes *The Malay Archipelago,* focusing on his voyage through the South Pacific, Borneo, Java and points south. The book will serve as a hopeful blueprint for generations of explorers and naturalists who wish to make their own discoveries in their travels at home and abroad.

1870-1882—Darwin's Later Years.

By now *Origin* has been in publication for more than a decade. The original rancor caused by Darwin (and Wallace) for the most part has died down. There are constant objections, but they aren't any more forceful, nor do they have the power to move natural selection off its scientific pedestal.

Then as today objections remain, and some are good and healthy scientific questioning of the theory. Others, like intelligent design, reek of theology and can be easily dismissed, at least by the majority of the scientific community if not the uneducated public.

As his eminence has grown, in the last ten years or so of Darwin's life there is an explosion of letter writing between him and the apostles. Most of it has an essence of gravitas mixed with humble elder statesman-like collaboration and advice. The same can be said about the number of letters written by Darwin to other scientists and family members alike.

1870 –Darwin writes to Wallace praising him for rebuking an article that attempted to refute and distort natural selection from a mathematical perspective. The well-written response by Wallace would appear in the November issue of the science journal *Nature.*

1871—Darwin publishes *Descent of Man,* in which he finally and publicly speculates on human evolution as it relates to biology and natural selection.

1871—Huxley publishes *Mr. Darwin's Critics,* a collection of essays taking on those who criticized Darwin based on the constant and growing evidence for natural selection

1871—Hooker writes to Darwin congratulating him on the publication of the *Descent of Man* and how (unlike *Origin* twelve years earlier) Darwin's work on evolution is being more easily accepted and is inspiring others.

1872—Asa Gray writes to Darwin and sends him along copies of his new book entitled *Botany for Young People—How Plants Behave.*

The book essentially is an accessible version for youth of Darwin's work. As Gray notes, *"I send you a little book, which may amuse you, in seeing your own science adapted to juvenile minds."*

1872—Hooker writes to Darwin about their mutual dislike of Sir Richard Owen, the anti-evolutionist and head of the British Museum who attempted to remove Hooker from his Directorship at Kew Gardens. Hooker writes, *"I am greatly gratified by your hatred of Owen for me.*[2] *How different our contempts are—I so despise him, that I feel I could afford to converse with him across a neighbour's table tomorrow—& yet I should be confoundedly angry if any friend of mine did so at the same table!"*

1872—Darwin writes to Wallace to thank him for again for forcefully defending their theory against hack scientists who for personal or other reasons reject natural selection. Wallace writes "a crushing" review of C. R. Bree's *"An exposition of fallacies in the hypothesis of Mr. Darwin."*

1872—Darwin completes and publishes *The Expression of the Emotions of Man and Animals.* The book focuses on evolutionary linked behavior between humans and non-human primates as well as other animals. While it does not sell out on first printing it is admirably accepted and serves as another brick in the evolutionary wall Darwin was building to link humans (as animals) to other animal groups. A second edition is published in 1890, eight years after his death, with an introduction by son, the botanist Francis Darwin.

1872/3—Wallace writes to Darwin telling him that he has written a very favorable review of *Expression* which will be published in 1873 in the *Quarterly Review of Science.*

1874 (June) Darwin writes to Asa Gray about Gray's article in *Nature.* *"I have now read your article in* Nature, *& the 2 last paragraphs were not included in the slip sent before…Everyone, I suppose occasionally thinks that he has worked in vain, & when one of these fits overtakes me, I will think of your article, & if that does not expel the evil spirit, I shall know that I am at the time a little bit insane, as we all are occasionally."*

1875—Draper publishes *History of the Conflict between Religion and Science,* the second of two worthy books which support Darwin's work in ecology and natural selection. This "knockout punch of a book" clearly demarks the birth of the scientific revolution and calls out religious philosophy as antithetical to human development and the modern understanding of the world. It also deeply supports Darwin's research and his later books post-*Origin.*

1876—Asa Gray publishes *Darwiniana: Essays and Reviews Pertaining to Darwinism.* The text is a tribute project, perhaps even a love letter to Darwin. Like many of the apostles, Gray expressly wanted to show his admiration and affection for Darwin. The goal of the text of course is to show respect but also memorialize the man and his ideas for future generations.

1877—Joseph Hooker travels to the United States and leads an expedition throughout the American west. Hooker is joined by fellow Darwin apostle, the influential American botanist Asa Gray, and together they travel and collect botanical specimens vigorously. By October, Hooker is back at Kew with more than a thousand specimens.

1878—Huxley publishes *Evolution in Biology,* a continued refutation of biblical ideas regarding special creation as well as a detailed examination of the biological basis for life based on evolutionary biology

1880—Darwin and Huxley exchange letters in early 1880. Darwin asks Huxley for advice as he (Darwin) has been accused of dishonesty by English novelist Samuel Butler.

1880 (April)—T. H. Huxley gives an address to the Royal Institute entitled, *The Coming of Age of the Origin of Species.* In yet another defense of Darwin's work, Huxley reads in grateful detail about *Origin,* its value back in 1859 and its relevance at close to the turn of the century. Like any good apostle, Huxley bears witness to the birth and controversy and correctness of *Origin*. He ends his talk by saying,

> Thus when, on the first day of October next, 'The Origin of Species' comes of age, the promise of its youth will be amply fulfilled; and we shall be prepared to congratulate the venerated author of the book, not only that the greatness of his achievement and its enduring influence upon the progress of knowledge have won him a place…but, still more, that,…he has lived long enough to outlast detraction and opposition, and to see the stone that the builders rejected become the head-stone of the corner.[247]

1880—Darwin pens a brief letter to his old friend Huxley thanking him for his defense and lecture, noting that Darwin had family members in the audience and that they, along with following newspaper reviews, gave Huxley's talk great favor.

1880 (November)—Darwin writes to Wallace praising him on his new book, *Island Life.* Darwin suggests that it's the best book written by Wallace in his career so far. The praise received is gratifying, as Wallace had always felt Darwin was an elder statesman of natural science.

1880—Darwin coauthors his second to last book with his son, Francis, an eminent and well-respected botanist. The book is entitled, *The Power of Movement in Plants.*

1880—Hooker writes to Darwin praising the near simultaneous publication of *Power of Movement of Plants* and Wallace's work in producing *Island Life.*

1881 (January)—Darwin writes to Huxley on the success for a memorial pension for Wallace. While both he and Huxley agreed to such an honorarium, Hooker was a holdout because of Wallace's belief in spiritualism. Darwin worked diligently throughout 1880 to secure the Wallace pension and in late 1880, Hooker begrudgingly agreed to sign on and so within months the pension was secured. Darwin will also write to Wallace in January 1881, and thank him for his contributions to science (as the basis for the pension).

1881—Darwin publishes his last book, *Introduction to Earthworms.* The book is published less than six months before his passing. While thought of as a lowly creature, the earthworm plays a significant role in ecology and horticulture. Praise for *Earthworms* came from many reviewers, including that of the Times of London.

1881—Hooker writes to his old and dear friend Darwin. Hooker laments the loss to death of many allies and colleagues and misses the times both he and Darwin shared together in their youth and through the trials and tribulations of each of their careers as naturalists.

1882—Darwin's apostle Dr. John William Draper passes away January 4, 1882. Draper's lecture in 1860 started the debate surrounding *Origin* and natural selection. His later books on the history of science, the European enlightenment, and the inherent conflict between science and religion all served to support Darwin's work in biology and natural history.

1882—(March)—T. H. Huxley writes to Darwin, worried about his health, and implores him to start using younger doctors from London. This is Huxley's last personal letter to Darwin. Darwin, still alive and always a gentleman, pens a letter in reply to Huxley on 27 March 1882, thanking him for his concern.

1882—Charles Darwin passes away on April 19. Although Darwin's life ends, his ideas are preserved for all eternity. Darwin was an extraordinary man who gave all humanity a key to how we can view our lives with dignity and understand our place in nature. How we maintain and manage this physical, emotional, scientific, and philosophical legacy is up to each of us.

1882—Huxley pens *Charles Darwin,* **as a memorial to his dearly departed friend**. He ends the article:

> None have fought better, and none have been more fortunate, than Charles Darwin. He found a great truth trodden underfoot, reviled by bigots, and ridiculed by all the world; he lived long enough to see it, chiefly by his own efforts, irrefragably established in science, inseparably incorporated with the common thoughts of men, and only hated and feared by those who would revile, but dare not. What shall a man desire more than this? Once more the image of Socrates rises unbidden, and the noble peroration of the "Apology" rings in our ears as if it were Charles Darwin's farewell.[248]

1888—American naturalist and botanist, Asa Gray, passes away on January 30, 1888. One of two "American apostles" and the only one actually born in the United States, Gray's advocacy for Darwin allowed for the acceptance of natural selection across the Atlantic. While Gray tried to find a plausible intersection between Darwin's theory and religious faith, he did so by insuring that Darwin's books were available in the United States.

1895—Darwin's bulldog, the brilliant and pugnacious Thomas Henry Huxley, passes away on June 29, 1895. Until the end of Darwin's life (and after) Huxley would be one of Darwin's fiercest defenders. He was a bulldog like no other, a fierce advocate for science unencumbered by religion, a campaigner for a host of social issues and also for modernity. Huxley lived his life not taking guff from any individual or institution. Perhaps that is why he was so feared in his life amongst his respective colleagues and detractors alike.

1911—Darwin's most revered friend and closest confidant, Joseph Dalton Hooker, passes away on December 10, 1911. Hooker served as a bridge for Darwin between people and institutions. It was Hooker who was Darwin's most trusted editor for *Origin* and for his later books as well. Although it was offered that he could be buried at Westminster Abbey next to Darwin, he chose while alive to be buried in the church cemetery that was steps away from the Royal Gardens at Kew.

1913—The last of the apostles passes on November 13, 1913. Co-discoverer of natural selection, Alfred Russel Wallace was a brilliant explorer and synthesizer of ideas. Without the social, educational, and financial advantages that Darwin had, he nonetheless managed to independently come to the same conclusions as Darwin with regard to natural selection. While his ideas regarding spirituality and phrenology do not mesh with modern science, Wallace remains a respected naturalist and deserves to be lauded for his advocacy work in favor of both Darwin and natural selection.

Some Final Thoughts on the Lives and Passions of Great Scientific Leaders

It is impossible to fully calculate the contributions of Darwin, Draper, Gray, Hooker, Huxley, and Wallace. How does one quantify all the hard work, all the dedication to science and knowledge, the time spent toiling, writing, sharing, debating, and infusing the world's understanding of how nature actually works—without magic—so that we can see all the beauty and dangers of the natural world.

We believe that it is deeply important to honor each of these men. We also believe it is almost impossible not to appreciate their work ethic, their contributions to science, their conflicts with the ideas and situations of the day and at time the problems they had on some occasions with each other. Each of the apostles who advocated for Darwin could not have imagined that they would be linking the rest of their lives and their careers to a sickly man, a person of self-doubt and guilt, a person whose anxieties surely created some if not many of his physical medical ailments.

But attached they had become. Why, you may ask? Well, for one let's remember that the best of science is the unbiased search for truth. Each of these men, Wallace especially, realized that natural selection was (and remains) a solid truth of nature. Place truth up against any theory or resulting opinion and, like Occam's Razor, the truth always wins. That's because while false ideas may temporarily daze us, truth is essentially timeless regardless if we accept it or not.

What then do we say in science classes today? How do we impress upon politicians the reality and urgency of the ideas contained in *Origin*? How do we impress this on the general public, who may be less interested in truth and more comfortable with several simple untruths? How do we do this when these untruths offer temporary comfort, even if this is at the expense of true knowledge?

Where do we go from here? Our simple advice is to look to the scientists in this book who risked their careers and in some cases almost their lives for the truth. If you can come to terms with this simple equation, you can become an advocate for science, for natural selection and indeed for modernity. The wonderful thing is it isn't even hard to do. Go to your local library. Get a book on science and start reading voraciously and happily. The truth you find on those pages make be a difficult first encounter, but like any exercise, the brain will respond and grow more resilient with every text you pick up, with every website that you find, and with every course you may seek to take to gain a closer approximation to the truth.

Doing this brain training certainly honors the scientists profiled in this book, but it does more than that. It makes you more knowledgeable and less open to errant ideas, falsehoods, and outright lies. You yourself can become a better arbiter of truth. As your knowledge grows, individuals and institutions that you trusted in the past may become less credible, and institutions that required trust from you may become less believable.

And in the end this is exactly what Darwin and the apostles stand for: breaking with the past, seeking new and more just ways to view nature and our human place within the structure of the natural world.

Each journey to greater understanding starts by placing your foot forward and opening you mind to possibilities outside your frame of current understanding. If you want to be good to yourself and those around as well as for future generations, then you *must* be ready to let go and jump into science with gusto.

Do this and you will do the memory, the achievements, and the names of Darwin, Huxley, Hooker, Gray, Draper, and Wallace very proud indeed.

CHAPTER 14

DRAWING A LINE: DARWIN AS APICAL FREETHOUGHT ANCESTOR

"What Galileo and Newton were to the seventeenth century, Darwin was to the nineteenth"

Bertrand Russell, *A History of Western Philosophy*

"Darwin made it possible to be an intellectually fulfilled atheist."

Richard Dawkins, *The Blind Watchmaker*

Modern day astronomers have developed several complex theories as to how and when the universe began. They conclude based on study, testing and evidence, that somewhere in the vicinity of fourteen billion years ago (give or take a few hundred thousand millennia) the universe exploded out of quantum nothingness. What came before is for theoretical physicists who study the mathematics of the natural world to uncover, debate and share with humanity. What came afterwards was first described in 1927 and then best expanded in 1931 by Catholic priest Georges Lemaître, and that of course is known as the *Big Bang*, thanks to astrophysicist Fred Hoyle.[249]

Since that auspicious eruption of everything from nothing (or perhaps something else?) the universe in which we all are born, live, and will eventually leave has been guided by physical laws and principles. These are ours to uncover and use for a host of possibilities. We big-brained primates of the genus *homo* get to enjoy this knowledge because of our own evolved curiosity. Because of our ability to be self-aware, conscious if you will, we have been searching through the natural world for many thousands of years to answer several fundamental questions about humanity's place in nature.

Those questions include different versions of, "Who are we?" "Why are we here?" and finally, "Where are we going?" In the centuries and generations before science could help lead us to approximate and measure our data-focused conclusions to these three questions, it was organized religion that held most of the approved answers. Pseudoscientific ideas and beliefs, like alchemy and astrology, also held sway over our sciences, beliefs, and imaginations.

Many of these religiously-based answers held because they came from authority and because the authorities concluded that they were timeless and unalterable.[250] Unless we sinned against the gods, humans could trust that the rivers would flow gently and disease would not harm our loved ones or crops, that mountain eruptions would be docile, and the land would not shake and tumble our homes to the ground.

Except that it never worked, and besides, human curiosity will out. Scientific truths, no matter how painful and upending to less verifiable ways of thinking remain the best way to know and understand our physical world. Scientific truths like the fact that the tides are controlled by gravity, or that the sun and not the earth is the center of our solar system, that our planet isn't a Frisbee-like flat disc hovering in the heavens, were cured by investigation, observation, and reason. Evidence that vaccines prevent disease and that humans evolved from a common ancestor shared by our closest ape relatives each in their own way upend un-evidenced conventional wisdom, much it based on wishful thinking and tradition.

Perhaps it is because knowledge of the natural world in general, and Darwin's theory of natural selection in particular, voids the need for deities and first causes which are based on mythology that they are perceived as such a danger to mythical and spiritual ideas. Darwin himself wrote about this apparent vulnerability we humans have for accepting special creation in his diaries. He well noted, "We can allow satellites, planets, suns, universe, nay whole systems of universes, to be governed by laws, but the smallest insect, we wish to be created at once by special act."[251] Darwin's ideas and his work have each served as catalysts for the modern development of the biological sciences but of the freethought movement as well.

Freethinkers and the Rise of Social Change:

Then as today, agitators cannot change the world unless at the same time they upset the current social order. Nineteenth century freethought activists frequently challenged the troika of orders of their time, which remain part of our time as well. These include the religious order that typically demands obedience to doctrine, the political order which maintains powerful alliances with connected families and their constituent groups in the legal and business world, and finally the social order that controls, maintains, and enhances the concentrated power of elites and those who aspire to become elite.

Their writings and activism confronted the social policies of the time which kept children out of school and in factories at all hours of the day and night. Focusing on their freethought ideas, they called into question the political system that stopped women from obtaining the right to vote. Working as socially conscious advocates, they demanded better treatment of ethnic minorities, who at worst were seen as sub-human and at best grist for the mill for ongoing hyper-industrialization. They also vigorously and openly questioned theological dogma which worked to maintain a self-appointed monopoly on wisdom.

These freethinking advocates were at once anti-war as much as they were pro-democracy, they were anti-establishment while being pro-freedom for all. They fought the status quo while also espousing pro-secular state justice where legal rights protected everyone regardless of economic status, class, age, or gender.

The nineteenth century freethinkers were themselves riding a wave of western liberal thought that had not been possible before the European philosophical reawakening from the Middle Ages. Their activism was fully possible because of the bedrock laid earlier by others who challenged the political stranglehold of the church-state relationship. In this case, the idea of links in a "great chain of being" are not meant as a biological pretext of advancing complexity, but one in which one generation takes the ideas of liberation and remakes them for new times. Revolutionary ideas are squarely at the heart of Darwin's work, as they were for his contemporaries, and as they were amongst all of the century's freethought activists.

The evolution of the freethought movement owes thanks to philosophers such as David Hume, Rene Descartes, Immanuel Kant, Jean-Jacques Rousseau, and Thomas Paine. Each held up a mirror to our ideals and choices. Each laid the ground for others to grow our materialistic philosophy and reality, eventually placing responsibility for our actions and existence into the hands of humanity rather than the omnipotent or omnipresent divine. Backed by the emergent sciences of modern geology under Charles Lyell and James Hutton, in astronomy under Nicolaus Copernicus and Galileo Galilei, amongst hundreds of mathematicians and other scientists in newly burgeoning fields, collectively they essentially helped reinvent science and our human connection to the natural world.

Darwin's Contribution to the Modern Anglo-American Freethought Movement

At this point in the book, it should go without saying that in the absence of Darwin and Wallace's theory of natural selection, observed and then codified in *Origin*, our very modernity would have been stalled or at least moderately delayed. Darwin and Wallace's theory helped to liberate the human mind from the confines of both bad science and religious-based thinking about our place in nature. It was also a catalyst for the further development of sciences like biology and chemistry, and it helped to bolster the nascent yet soon to be burgeoning freethought movement.

Perhaps the modern freethought movement, itself a separate species grown out of the philosophy of humanism and the idea of observable material nature, is itself connected to our progressive social evolution. As an unintended consequence of Darwin's work, *Origin* offered the freethought movement necessary ammunition for the growth of social and secular justice. With more than 150 years of distance, we can certainly see how *Origin* served as a pivot point for humanist and freethought social activists and scholars. From this same vantage point of time, we can consider Darwin's work a catalyst for truth then as it is today.

British and American Freethinkers and Darwin

In late nineteenth century America, the freethought movement was characterized by the speeches, writings and work of activists such as Robert Ingersoll and Elizabeth Cady Stanton.[252][253] In Europe, men and women like Charles Bradlaugh, George Foote, Annie Besant,[254][255] and many others led the way. Some were already convinced agnostics or full-on atheists, yet each saw their support of liberation movements as allied to the liberation equally gained by understanding the natural world. Indeed, Darwin's work was central to firmly ground our human place within the material world. So it remains a fine ally to the humanist and secular cause.

The Great Agnostic: Robert Ingersoll

Perhaps no other nineteenth century freethinker could crystalize the importance of Darwin's contributions to the freethought movement better than Robert Ingersoll.. He sadly was almost forgotten but has gained new esteem thanks to the excellent work of author Susan Jacoby. In his speeches and writings, Ingersoll lauded Darwin at every turn. Ingersoll ensured that if people had not read Darwin's work, they would at least know how Darwin's books and theories were responsible for our modernity.

Ingersoll saw Darwin as serving as *the* apex to a world centered on scientific and verifiable reality rather than on theological observation and subjectivity. He writes, "This century will be called Darwin's century. He was one of the greatest men who ever touched this globe. He (Darwin) has explained more of the phenomena of life than all of the religious teachers."[256] One can certainly see Ingersoll's exaltation of Darwin as well as his veneration of Darwin as both a man and scientist.

But one quote alone, no matter the veracity, could not stand as full evidence of Ingersoll's acceptance and admiration of Darwin. Ingersoll's writing is indeed littered with adulation and praise for the naturalist. Here is Ingersoll taking on religious dogma in the face of Darwin's theories:

The Church teaches that man was created perfect, and that for six thousand years he has degenerated. Darwin demonstrated the falsity of this dogma. He shows that man has for thousands of ages steadily advanced; that the Garden of Eden is an ignorant myth; that the doctrine of original sin has no foundation in fact; that the atonement is an absurdity; that the serpent did not tempt, and that man did not fall.[257]

In the above quote, Ingersoll wholeheartedly pulls down the curtain of false reality of competing religious magisteria. These were and are the poor alternative facts presented by religion, organized faith, and dogma. Ingersoll in that same paragraph simultaneously lauds Darwin for his work, as a key to giving him the strength to leave more ancient ways of knowing behind in favor of material truth based on scientific observation and verifiable evidence, beyond our senses.

In his deeply passionate lecture and the later pamphlet of the same title, *Some Mistakes of Moses*, Ingersoll lives up to the moniker as the "Great Agnostic." He provides an eloquent takedown of clerical ignorance. He writes:

As a rule, theologians know nothing of this world, and far less of the next; but they have the power of stating the most absurd propositions with faces solemn as stupidity touched by fear...It is part of their business to malign and vilify the Voltaires, Humes, Paines, Humboldts, Tyndalls, Haeckels, Darwins, Spencers and Drapers...They are, for the most part, engaged in poisoning the minds of the young, prejudicing children against science, teaching the astronomy and geology of the bible, and inducing all to desert the sublime standard of reason.[258]

Now firmly caught in the tractor beam of Ingersoll's respect for Darwin's work, let's look at perhaps the coup de grace of Ingersoll's simultaneous ire towards religion and deep respect for Darwin. Ingersoll writes,

Charles Darwin destroyed the foundation of orthodox Christianity. There is nothing left but faith in what we know could not and did not happen. Religion and science are enemies. One is a superstition the other is fact. One rests upon the false, the other upon the true. One is the result of fear and faith, the other of investigation and reason.[259]

Of course this spotlight and adulation was something that Darwin himself feared during his lifetime. It is the reason why he took more than twenty years to publish *Origin*. It would be his worst fears come true that others would rightfully use his observations and research to overturn his wife's deep faith, albeit a faith that he himself had abandoned.

Elizabeth Cady Stanton, Freethought, and the Suffrage Movement

Elizabeth Cady Stanton is certainly remembered for her many decades long work inside the woman's suffrage movement. However other major freethinkers, at least in America, remain outside our memory and fade into the mist of the historical record and who will be remembered here.

For Stanton and other leading nineteenth century woman's suffrage movement leaders, Darwin's work bolstered their idea that women held the keys to evolutionary success rather than men. They took what Darwin offered and showed that females of every species, once in control of their reproductive lives, could manage when and even if they would bear offspring. These feminists suggested that should women in "civilization" have the same ownership of their bodies that animals had in the wild, that it would empower women around the globe.

The feminists concluded that animals in nature chose the best mates based on how the male would dance, smell, or show plumage and it was a female's choice during mating to select the individual most fancied. They accepted that if a human woman had the right to choose her time of pregnancy she therefore would be the most powerful force in shaping our human future. A woman's right to choose to delay conception or not conceive at all therefore was not only an evolutionary determining factor in evolutionary success, but stood as a human political right as well. The demand for social and reproductive rights of these freethinking suffragettes grew louder and into large-scale political movement as the clock turned closer to the twentieth century.

Although Charles Darwin's writings were used to support the suffrage movement, and in fact Darwin himself did write that female choice of mates was an important factor in evolution, Darwin himself was not in any way allied to the feminist cause. In fact, inside the publication of *Descent of Man* in 1871, Darwin expressly writes that evolution endowed men with greater intellectual abilities.[260] Darwin, a man ahead of his time in so many ways scientifically, was also a man of his time as well. Darwin felt that men should paternally care for women, and that as a lesser gender women resided in a sort of halfway stage, between babies and adult men.

This very real bifurcation showed that Darwin was as much a scientific revolutionary as he was invested in nineteenth century patriarchal cultural status quo. However, knowing Darwin's inquisitive nature and trust of scientific evidence, perhaps had he lived longer his views on suffrage would have changed. However, it should be noted, not all men of Darwin's ilk were so paternalistic or felt a woman's best choice was to abdicate her will and body because of her gender. As a for instance, the apostle bulldog T. H. Huxley, a man who rose from poverty, had in fact worked for the cause of woman's suffrage and human rights throughout his lifetime, and Alfred Russel Wallace likewise vocally supported women's rights, equality and suffrage.

Nevertheless, Stanton took from Darwin's work the essential scientific materialism that placed all humans in the rational world rather than within any metaphysical realm. In her speech, *The Pleasures of Age,* given in 1885,[261] she spoke of the naturalist Alexander von Humboldt and his view of both reality and all modern science that was governed by physical laws that served to enliven our understanding and how such knowledge should lead to greater human fairness. She shared, "…all things are governed by laws. And this law is immutable in the moral as in the material world…We shall find that the keynote to our human relations is love and the grand chorus is equality."

Darwin's personal views on gender relations were certainly antithetical to both the American and British suffrage movements. But his observations and writings on nature did provide scientific evidence for social and freethought activists. Stanton, like many of the men and women who fought for the suffrage movement, was able to proclaim and fight for a world where all people were indeed created equal based not only on progressive philosophy but on scientific truth as well.

Stanton, a suffrage advocate and freethinker till the day she died, penned one of her last letters to the *New York Evening Post* about the Bible being rejected in public schools. For decades in her speeches and writings she viewed the Bible as a major impediment not only for freedom for women but freedom to think scientifically and freely as well. Two weeks before her death she wrote, "I suggest that inasmuch as the Bible degrades woman, and in innumerable passages teaches her absolute subjugation to man in all relations…in its present form, should be taken from schools, and from the rising generation of boys, as it teaches lessons of disrespect for the mothers of the race."[262]

Charles Darwin, while obtaining a degree in theology, had also forsaken his religious faith because of his research and writings. While he did so painfully, both he and Cady Stanton saw the Bible as fully in conflict with material truth. As such they were naturally allied even if they diverged on the social issues of the day related to women's social and political rights. Darwin was clearly in the wrong. Cady Stanton lived well past the naturalist but sadly, not long enough to see the Nineteenth Amendment to the Constitution come to pass in 1920, which gave women the right to vote in the United States.

Forgotten American Freethought Heros: D.M. Bennett and Andrew White
D.M. Bennett

Equally courageous (if not as dangerous to the detractors of early freethought) was **DeRobigne Mortimer (D. M.) Bennett**. Bennett was the founding publisher of the publication *The Truth Seeker,* the first major late nineteenth century American freethought newspaper.[263] This tome was dedicated to examining a rational godless universe. It also simultaneously exposed the fallacies and detriments of faith-based ways of life and their allied philosophies.

D. M. Bennett's respect for Darwin was wide as it was very deep. As the editor and publisher of *The Truth Seeker,* Bennett included articles and think pieces about Darwin's work frequently in the publication. As publishers, America's Bennett and Great Britain's Bradlaugh were much like two strands of freethinker DNA, separated not so much in thought but by continent. In fact, Bennett would republish an article from Bradlaugh's *National Reformer* publication that included a key interview by E. B. Aveling with Darwin, in which Darwin is quoted noting his beliefs: "I am with you in thought, but I should prefer the word 'Agnostic' rather than 'Atheist'…I never gave up on Christianity until I was forty years of age…It is not supported by evidence." [264]

First published in 1873 and still in publication today, *The Truth*

Seeker served as a major platform to share freethought ideas and serve as an outlet for popularizing Darwin's research. The newspaper serialized both *On the Origin of Species* and the *Descent of Man*. Alone these are incredibly important works of science. But taken together they are Darwin's one-two punch that not only placed humans well within the material realm of evolutionary biology but also distinguished science as a sincere and primary mediator for understanding nature and truth.

Just as apostle Asa Gray would secure Darwin's publishing rights in the United States, the *Truth Seeker* sought to serve as an outlet for Darwin's observations and research as well. With a subscription list of more than fifty thousand readers, including such notable authors and intellectuals as Mark Twain, Robert Ingersoll, and Clarence Darrow, the paper was as much acclaimed for its content and it was for its central purpose, to push material reality over superstition and theology.

Darwin and Bennett would both die in 1882, Darwin in April at seventy-three and Bennett at the youthful age of just sixty-four in December. The ranking *Truth Seeker* editor at the time, Eugene MacDonald, wrote a scalding editorial essentially condemning the burial of Darwin inside Westminster Abbey, since only a few years earlier the Anglican Church had condemned Darwin's work as an affront to religion.

Years after Darwin and Bennett's deaths, the *Truth Seeker* would continue to publish pro-Darwin articles and titles, including *Moses or Darwin: A school problem for all friends of truth and progress*. During the Scopes Monkey Trial, the paper intensely covered the legal proceedings, focusing on the abject anti-Darwinism, the social and religious implications of the trial and the danger religious belief could result in a slide backwards for science and our human understanding of nature.

A.D. White

Andrew Dickson White's contribution to the American freethought movement would serve as a starting point for many later leaders. White's treatise about the problems and conflicts between religion and science was equivalent in impact to Draper's own work on the same subject just decades earlier. A very cautious and studious researcher like Darwin, White spent years observing, editing, and refining his work in the same cautious way that Darwin would come to write *Origin*.

The reasons for that caution might have been different, simply because Darwin was so fearful of the impact his work would have on his family and English society. Meanwhile, White was much more worldly in his responsibilities and just too busy in his diplomatic and later academic career to just focus on one area for any length of time to be able to publish any sooner.

Perhaps not so ironically, White also had a lot in common with another Darwin apostle who was also a believer. While he rejected religious doctrine in favor of the use and methods of science to understand the universe, White, like Asa Gray, was a believer in the divine. White, though, refused to accept theology as leading to understanding any realm of nature, and he viewed theology as a form of human corruption.

He notes, "Science, though it has evidently conquered Dogmatic Theology based on biblical tests and ancient modes of thought will go hand in hand in religion."[265] As a champion for academic freedom at Cornell, in *Warfare*, White would write about the forces against and the ultimate contributions of Charles Darwin:

> All opposition had availed nothing; Darwin's work and fame were secure. As men looked back over his beautiful life—simple, honest, tolerant, kindly—and thought upon his great labors in the search for truth, all the attacks faded into nothingness.[266]

It is clear that the White saw the deep conflict between religion and science, even as he tried to separate the divine as a structured force for good from both theology and scientific evidence. Although he was not a scientist, White knew that science itself could not speculate on the metaphysics of any omnipotent being without physical evidence that could be tested. This issue of evidence remains with us today and still eludes skeptics and the faithful alike, although for believers in most religious traditions, no such bar is required to pass. Belief is based on the acceptance of faith and not evidence of fact.

As an outside observer with great influence, White was indeed captivated by the efforts of science to broadcast a greater truth to humanity than humanity itself was able to create through the use of religion. It is perhaps for this reason that White was such an admirer of Darwin. White saw the great naturalist and his efforts to discover the rational way nature impacts all species as a sign that humans were gaining a greater truth about the nature of reality itself.

White would write, "The inquiry into Nature having thus been pursued nearly two thousand years theologically, we find by the middle of the sixteenth century some promising beginnings of a different method of inquiry…The method which seeks not plausibility but facts."[267] As such, in America and abroad, even though White might be considered a Deist, his contributions to the freethought movement as well as his support of Darwinian theory makes him an MVP in diplomacy, secular higher education, academic freedom and the improvement of scientific understanding.

White is also credited for helping to found Cornell University, in New York State. He demanded that Cornell be founded as a secular institution. This Ivy League University was the academic incubator of such esteemed scientists as Carl Sagan, and students such as astronaut Mae Carol Jamison, science popularizer Bill Nye, and astronomy writer Alan MacRobert. Cornell University was also the academic home of atheist provocateur Bill Maher, the writer/director of the anti-theistic film *Religulous*.

As diplomat for the United States during the American Civil War, White helped secure promises from both the governments of England and France not to assist the breakaway southern states. White spent decades of his life refining his basic thesis in which he saw religion as antithetical to truth. A full decade after the publication of *Origin* in 1859, White gave a formational lecture on the conflict between religion and science. This later culminated in his major work published in 1896, *A History of the War of Science with the Theology of Christendom*.

White's *War of Science* reads much like that of Darwin's apostle John William Draper's book, *History of the Conflict Between Religion and Science*, published in 1874. Both authors saw religion as a grand inhibitor. They each wrote that theology was not only in conflict with scientific truth but that organized religion actively worked as a deeply threatening political and social impediment to both scientific inquiry and modernity. White's work echoed not only Draper's main thesis but it also bolstered the work and writings of fellow freethinkers, especially when it came to the schism between religious ways and scientific ways of knowing, thinking about and understanding the universe.

The 19th Century British Freethought Movement: Bradlaugh, Besant, *et al.*
Charles Bradlaugh

In England, Annie Besant and Charles Bradlaugh were each firebrands and activists for the freethought movement. Later in their careers, they would make a formidable team that confronted religious morality both at the state level as well as in the real lives of everyday British citizens. Born in 1833, Bradlaugh by 1866 was the chief editor of the secular newspaper, the *National Reformer*, and was the founder of Britain's National Secular Society.

Bradlaugh was also a politician. First elected to Parliament in 1880, he went to jail for refusing to take the religiously-based "Oath of Allegiance" all MPs were required to proclaim before taking their legislative seat. This began an eight-year journey whereby Bradlaugh was disallowed from voting and fined a considerable sum for voting without officially sitting in Parliament. His refusal also led to his imprisonment.

Today, Bradlaugh might be considered a conscientious objector, but in his own time, he was seen as a heretic and anathema. It was not until 1888, almost a decade after his first election win, that Bradlaugh won the legal right to affirm his allegiance in a secular statement rather than take a religious oath.

Britain's Annie Besant, like Emma Goldman and Margaret Sanger in the United States, were considered radical and influential activists who fought for women's rights and against the immense patriarchy of English social and religious laws. At the height of her freethought activism Besant was a member of the Fabian Society, a British progressive organization that sought to democratize and alleviate the suffering of the working class. She and many female socialists of the time saw this suffering entwined with woman's suffrage and believed it was caused in part by rampant industrialization and the lack of laws that would protect worker rights as well as woman and child labor.

Later in life Besant would break away from the progressives within the Fabian Society and also the secular skeptics of her youth. This occurred essentially because of a chance encounter with the spiritualist philosopher and author Helena Blavatsky. Besant was so moved by Blavatsky's writings that she would forsake secularism for the rest of her life and would advocate for and firmly champion the mystical trappings of Theosophy.

Theosophy can best be described as a form of spiritual journey where the believer attempts to gain knowledge by seeking the hidden meaning of ancient spirit realms. This journey would, for those who believe, lead to greater awareness and revelation. Such practice is intended to lead the faithful to access the divine and ultimately find personal salvation. The idea of accessing the deeper meanings of life's truths through discovering unseen magical worlds is itself ancient. Today, followers of Theosophy are alive and well and can found around the globe and in almost every nation.

At their most progressive, Besant and Bradlaugh were partnered "radicals" who worked together to improve the lives of women. This activism took its sharpest form by their offering birth control information to the masses. Such activity was then, as it is now in certain states in the United States and around the world, tantamount to the highest sin. The two republished American Charles Knowlton's 1832 (almost forgotten) pamphlet, *Fruits of Philosophy*, which essentially offered married couples advice on how to avoid pregnancy. Knowlton himself was fined and spent several months in prison for authoring and distributing what was considered a vile tome.

Forty-five years later however, with little change in terms of the conservative elite prescribing and manipulating public morals, Besant and Bradlaugh published a revised version of Knowlton's work with additional detailed medical notes. They subsequently faced the same charge of abuse of public morals by the offended church and the conservative political class. Both were fined and arrested, which was indeed the outcome they wanted to prove a point about the free press. In their 1877 trial, the pair was eventually found guilty of disseminating a morally corrupt publication yet not of disseminating the materials with intent to harm the public, a strange twist in juror interpretation of British law.

Because of this split, their legal appeals left both defendants free pending the outcome of the further litigation. During this time they continued to press and make the pamphlet available to the general public. This infuriated both the religious elites and the legal system, which demanded that the defendants stop offering copies of Knowlton's work to the public during the appeal process. Both Bradlaugh and Besant promptly ignored this demand and would eventually sell more than 125,000 copies thanks to the notoriety of the case that they eventually won on appeal.

Bradlaugh even wrote to Charles Darwin about their plight. In a June 5, 1877 letter, he asked Darwin for his support with the intention of subpoenaing him to speak as a friendly witness.[268] Darwin by now was an internationally famous figure. Although they attempted to summon Darwin to their defense regarding family planning, Darwin would not in any way publicly or privately support Bradlaugh and Besant. Darwin was firmly against a woman's right to choose the timing of her pregnancy, believing that such an option would ultimately destroy the family and undermine selection.[269]

Just as Darwin could not support the suffrage movement he also could not support birth control. This blind spot is something that may tarnish his memory, but not his ultimate legacy and contributions to science or our understanding of nature. Darwin had his biases, as do we all. However, it is indeed possible that if Darwin had chosen to help defend Bradlaugh and Besant, they might not have been judged so harshly and thought to be a danger to British society for their progressive views.

Bradlaugh and Besant would go on to publish hundreds of freethought pamphlets that extolled nonbelief as well as Darwin's scientific views of nature. In *The Atheistic Platform,* an 1884 collection of freethought authors and essays published under the auspices of their The Freethought Publishing Company, Darwin's work took the stage in many of the book's chapters. In one essay titled, *"Is Darwin Atheistic?"* by Charles Cockbill Cattell, the author writes:

> Darwin's anticipation of the judgment passed upon his views has been more than realised. The great objection to his view is commonly expressed in the words—what it leads to. There can be no doubt that it leads to the assumption of natural instead of supernatural causes...Those who do not wish to relinquish their notion of the supernatural producing, sustaining, and guiding the natural had better leave Darwin alone.[270]

Cattell, a dedicated secularist, is certainly a lesser-known freethinker in modern circles today. But in his time his reputation as agitator and author was well respected. Perhaps he will be best remembered as not only a supporter of Darwin's work, but for his 1878 small volume, *The Martyrs of Progress.* This book included many historical biographies of the philosophers and scientists whose writings helped support the growth of western knowledge and progressive ideas concerning the natural world. Cattell was also a scientist and geologist who wrote about natural history and biology.

In this same publication, Bradlaugh would summon the names of both Darwin and his bulldog, T. H. Huxley. He writes, "I told you from the outset that I come to these conclusions by way of science. From science, especially your Darwinian science, you can learn so much. You that are students of Darwin, and have learned something of his views and of his great truths, will know what I mean by this idea of changing the environment, the surroundings, as well as changing the individual."[271] Even though there may have been animosity between Bradlaugh and Darwin, it is clear the Bradlaugh looked beyond any personal indignation in support of the greater truth of Darwin's works.

In another essay in the book, Bradlaugh writes about the dangers of errant capitalism and the idea of social economic competition, which he views with disdain. He juxtaposes such competition with the open sharing of important ideas of natural selection within the natural sciences. While an anti-elitist, he acknowledges the human liberation provided by the scientific elite and exclaims, "How is it all your great scientific work has to be done by men of means or holding sinecures?—your Darwins, your Huxleys—all these men who do all your best scientific work, but do it in no spirit of competition."

Essentially Bradlaugh draws a comparison between industrial capitalism, which he sees as favoring the few wealthy and that of scientific truth, which he believes democratically favors everyone. In an 1884 debate in which Bradlaugh participates, again Darwin and other scientists are lauded for their selflessness toward the free expression of knowledge:

I appeal, sir, to higher motives that have governed mankind, no; to low personal greed and profit which leads each man to strive to cut others throats for personal advantage. I have yet to learn that Newton or Simpson or Darwin or Faraday worked as they worked for the sake of individual greed or advantage. They did not; they worked for the good of the human race.[272]

In 1883, one year after Darwin's death, Bradlaugh's press would also publish the slight pamphlet, *The Religious Views of Charles Darwin,* written by the biologist and freethinker Edward Aveling. This work implored readers to remember Darwin's views of life as well as his fall from the theistic tree. This was a very real concern since Darwin was buried in Westminster Abbey and religionists would attempt to rewrite Darwin's life to make his views more favorable towards religion. Aveling writes, "they slandered him when living, not without protest from us; and now that he is dead and even evil is spoken of him, our voices must be raised."[273] As if owning his resting place meant that the church could own his history and memory as well—something freethinkers at the time and now rightfully challenge.

In an effort to continuously support the ideas of Darwin for the sake of the freethought movement, Bradlaugh's press produced another 1884 pamphlet entitled, *The Absence of Design in Nature.* This was essentially a republished 1883 lecture by American chemist H. D. Garrison, in which the author notes, "The doctrine of evolution—natural selection—the survival of the fittest, explains all in a most satisfactory manner. Evolution is, therefore, the designing hand...Those who refuse to accept the doctrine of evolution, because all the steps and stages in the evolution of animals and plants have not been observed...(are) rejecting the evidence of an army of witnesses."[274] Garrison notes later in his lecture:

> Since all the adaptation observed in nature is fully and rationally accounted for by the theory of evolution—indeed, we might say, is required by that theory—it is plainly a violation of the fundamental laws of human reason to attempt to explain these relations by invoking miraculous agency—a cause unknown to science, and of the existence of which no proof can be given in this age.[275]

Certainly in the nineteenth century, we see the rejection of religious faith and a turn towards scientific modernity in many of the freethought publications of the day. This was indeed made possible by those few heathens and radicals who paid a high price for their efforts. They not only supported Darwin and the natural sciences, but worked for the free and open access to information and the claims made openly to challenge white male European domination. All this work was done for the sake of wider and more inclusive human and civil rights, and a more just and secular world.

Annie Besant

For the once freethought activist turned mystic, Annie Besant's invocation of Darwin's name and work had both utilitarian and self-preservation purpose. There are numerous occasions when Besant used the naturalist's writings and research to make points regarding the suffrage and secular movements. She did so to prove her and Bradlaugh's innocence regarding the "unfit nature" of the materials they were distributing regarding female reproductive health.

In her 1877 publication, *The Law of Population. Its Consequences and its Bearing upon Human Conduct and Morals,* Besant calls upon Darwin's work from *Origin* and *Descent of Man* five times in the forty-seven page publication to justify her ideas and conclusions. She notes, "Darwin in his *Origin of Species* [said] there is no exception to the rule that every organic being naturally increases at so high a rate that, if not destroyed, the earth would soon be covered by the progeny of a single pair."[276]And while she views the harshness of natural selection as being without moral guidance, Besant concludes that because humans are ethical and moral animals, that Darwin's work must be seen as only part of humankind's success.

Besant did not have all the research we have today on the role empathy plays in our biological and social lives, but it is clear that there is an evolutionary predisposition for empathy wound tightly into our DNA. This has, in part, led to human success as a species. In her remarks about nature being out of balance, she again quotes Darwin to justify her ideas, "This decrease of natural checks to population...has a dark side. Darwin has remarked: 'Lighten any check, mitigate the destruction ever so little, and the number of the species will almost instantaneously increase to any amount.'"277 Even without his support of the suffrage movement his value to Besant was clearly evident.

Although Darwin did not come to Bradlaugh and Besant's aid during their trial his name appears in the court transcript in support of their defense many times. Mainly, Darwin's research is used to suggest that survival of the fittest targets population imbalance. Under pressure, natural selection typically wipes out the majority members of a species. Darwin does write that all populations are under such ecological pressure. At trial, Besant invokes Darwin: "in his great work *Origin of Species* he says, A struggle for existence inevitably follows, from the high rate at which all organic beings tend to increase."

By some accounts, Darwin was very upset and angry that the pair were trying to wrest him from his Down House fortress for their defense, even prior to knowing where he stood on the issue of family planning. Being a paternalistic Malthusian, Darwin saw the idea of a woman's right to choose the timing of her pregnancy as dangerous for two reasons. He worried that if women had such a choice it would have implications on birth rates. More women choosing the time of their pregnancy he felt would be detrimental for our species success, a reversal if you will of natural selection.

Darwin also believed that women needed a man's guiding support to ensure their safety and ability to produce offspring. Therefore such options regarding conception would upend the central gender roles within the family. To be specific, in his reply to Bradlaugh, Darwin writes, "...artificial checks to the natural rate of human increase are very undesirable and that the use of artificial means to prevent conception would soon destroy chastity and, ultimately, the family."

Besant and Bradlaugh concluded from Darwin's writings that empowered females who choose when they become pregnant can thus support our species' ability to maintain a constant balance. They advocate that such empowerment therefore can avoid the violence of hunger, famine and the wasting of whole human populations. Without looking into a crystal ball, their thesis is validated. One needs only to look at unchecked human population growth today and see a high correlation between a lack of birth control and poverty, food shortages, abuse, disease, homelessness, high infant mortality, and lack of health-care and education.

Later British Freethinkers: George William Foote and Chapman Cohen
George Foote

George Foote can best be described as a young protégé of Bradlaugh who deeply respected the elder secular leader. He too spent time in prison on charges of publishing and distributing materials "unfit" for reader consumption. This really meant that his atheist writings and his founding of the publication *The Freethinker* was too much for the religious and non-secular political authorities of the day.

Foote had a deep admiration for Darwin as well. Seven years after Darwin's passing, Foote wrote admirably about the late naturalist. In a pamphlet written in 1889, Foote compared Darwin to Newton: "In a certain sense, however, Darwin's achievements are more remarkable, because they profoundly affect our notions of man's position and destiny in the universe." The use of pamphlets exploded in the nineteenth century and the freethought movement used cheap printing to their advantage as a way to share ideas with larger numbers of people. The best way to think about pamphlets is to see them as the pre-Internet blogs of today. That is how quick and potentially impactful the messaging was back in Darwin's time.

Upon his conviction for printing alleged blasphemous material, Foote spent a year in jail working in a hard labor camp. After his release, his dues paid, he went on for years as a tireless advocate for atheist, secular and freethought causes. Foote lived into the twentieth century, passing away in 1915, yet he never forgot the price he paid or the feelings of injustice he was served for his ideas, writings, and activism.

So Charles Darwin's name would be ensconced in another famous British trial of yet another then famous nineteenth century freethinker. As noted earlier, George Foote did not have the same beneficial outcome of freedom until appeal as Bradlaugh or Besant. Foote would spend a year doing hard labor on the charge of blasphemy because of his freethought and atheistic writings.

In his 1883 trial, Foote sought to show that the ongoing persecution of freethought and science ideas was a poor and losing effort on the part of the nation's legal system. He notes, "Rev. 'Dr.' Wainwright at its head, which has announced in the 'Times' its intention of prosecuting the works of Messrs. Huxley, Tyndall, Darwin and Spencer…This may appear ridiculous, but the ridiculousness is the fault of the law, not the men."[278] Such frustration at the idea of both the church and legal system exerting so much pressure on fellow citizens to abandon their ideas is palpable in Foote's opening defense.

Sadly though, the pressure and censorship is a tradition still in practice today. The purpose, then as now, is to maintain political power by controlling the language, thoughts, and conversation of those governed by the state. Both the International Humanist and Ethical Union's annual freethought report and various United Nations publications focus on state and non-state actors that harm writers, toss them into prison, or violently harass and even kill those who would question faith.

Foote's most substantial support of Darwin came long after his release from prison. In 1889, Foote penned *Darwin on God,* a biographical sketch of Darwin that focuses on his eventual loss of religious faith. Foote wrote;

My object is to show the general reader what were Darwin's views on religion, and, as far as possible, trace the growth of those views in his time. I desire to point out, in particular, how he thought the lead ideas of theology were affected by the doctrine of evolution. Further, I wish to prove that there is no essential difference between his Agnosticism and what has always been taught as Atheism.[279]

The book opens with a review of the Darwin pedigree, noting that Charles's grandfather Erasmus and father Robert were both men who accepted their Anglican faith as well aspects of Unitarianism, even if at times they might have had self-doubts or were even skeptical about aspects of the dogma and doctrine. His book also recounts Darwin's early acceptance of faith and his later graduation in 1831 from Cambridge University with a Bachelor's Degree in Theology.

Like the dual mechanisms of evolution, the slow accretion of traits and the gradual change over long stretches of time, Darwin's ultimate agnosticism came to him later in life. Foote eloquently notes these subtle but important changes in Darwin's thinking. He writes, "It is therefore obvious that Darwin doubted Christianity at the age of thirty, abandoned it before the age of forty, and remained a Deist until the age of fifty. The publication of *Origin of Species* may be taken as marking the commencement of his third and last mental epoch."[280]

Foote goes on to write, "The philosophy of evolution took possession of his mind, and gradually expelled both the belief in God and the belief in immortality." [281]Foote's substantiated his claims of Darwin's disbelief after long review of his later writings and conversations with colleagues and family members. These served to verify his loss of religious faith and lack of repentance even upon death. Foote, like any good historian or ethnographer, also interviewed those closest to Darwin. After all this research Foote's conclusions regarding Darwin's agnosticism remained firm.

Chapman Cohen

Perhaps the youngest of that generation's British freethought orators and advocates is the oft remembered **Chapman Cohen**.[282] Like Foote, in his time, Cohen was a fighter for the secular movement and a prolific author. Born in 1868, he lived through half of the twentieth century. He was an in-demand speaker, giving more than two hundred lectures in a year, and frequent contributor to the *Freethinker*. In1898, Cohen became the publication's editor. He would also come to sit in the same leadership chair as Foote as the president of the National Secular Society. Cohen was a prolific and sincere freethinker, a life-long advocate for science and Darwin's work, and a man dedicated to both skepticism and secularism.

Cohen's intellectual energies were never in short supply and his writing frequently focused on the freethought issues of the day. As an author and advocate who spanned both the nineteenth and twentieth centuries, his was a special link in the chain between generations of secular advocates. Case in point, in 1896 he penned *An outline of Evolutionary Ethics*; in 1897 he wrote *Evolution and Christianity.*

Both of these works strongly support Darwin's research and scientific contributions to our understanding of nature and the development of human values. In *Evolutionary Ethics*, Cohen writes, "The object of the present essay is disclosed in the title; it is that of presenting, in as few words as possible, an outline of a system of ethics based upon the doctrine of Evolution."[283] Here, Cohen draws a straight line between our evolution and the ongoing development of humanistic values that keeps our species motivated towards doing good.

Conversely, Cohen views theology as static and unsympathetic to the plight of humanity. He writes, "The great weakness of all theological and metaphysical systems of morals, is, that they take man as he is, without reference to his post history or evolution, and proceed to frame rules for this future guidance."

And in *Evolution and Christianity,* Cohen pens the following take down of a creator: "But if it was God's purpose that only the fit should survive, why create the unfit? If the wisdom of God is shown in the structure of animals that survive, must there not be a sad lack of wisdom in the immense number that succumb?"[284] Clearly there is no otherworldly hand in our existence. Darwin came to know this fact and it has been reinforced both strongly and diligently by Chapman Cohen and by so many of other freethinkers, ancient and modern, all for the betterment of humankind.

In their time, these "radicals" moved mountains to promote educational, labor, social and economic justice. They wrote about and supported the legal protection of minorities and spoke openly about wage and work injustices hurting the poor. They dared the politicians of the time to write laws protecting both children and women from predation and socio-economic violence.

Darwin Acknowledges and Responds to the Activists

For all the freethought activism happening in England and the United States, Charles Darwin remained rather subdued and secluded from the winds of social change. Of course he was a deeply focused scientist who never really even took a stand in his own defense once *Origin* was published. So it may have been hard for him to work toward other forms of intellectual openness, physical freedom, and social liberation that the world had taken to faster than ever before.

Such advocacy, if Darwin decided to lend his voice, would have taken him away from his research. To Darwin such loss of focus would have been a cardinal sin. And coupled with his famously well-known bouts of real and stress-related illness, Darwin would not have had time to complete his research and be a freethought activist. Then as today, there are many ways to make a difference and for Darwin, his sentiments and strengths fit nicely into the "thought leader" rather than boots-on-the-ground activist or open checkbook.

However much Darwin chose to balkanize his time, personal connections and living arrangements he certainly was not by any means a loner. He had his wife, his children, his staff, and his colleagues, who all served to be at his side and at times to either challenge or entertain him. Darwin also had his books and correspondence. Darwin was an avid letter writer and this skill helped him to connect with scientists, students, as well as with friends and admirers alike in the United Kingdom and from around the globe. In today's world I can only imagine that Darwin's Facebook and Twitter accounts would be oversubscribed.

Darwin may not have been an advocate for the suffrage movement or of a woman's right to choose the timing of her pregnancy. But he was sorrowful for other forms of human bondage such as abject poverty and slavery. Darwin was anti-slavery and wrote about the evils of slavery as well.

It was in fact Darwin's in-laws, the wealthy and connected Wedgewoods, that brought tremendous political pressure on Parliament and helped create the final legislation that would ban British slavery in 1833. Darwin saw slavery first-hand during the *Voyage of the Beagle* trip, and it left him feeling devastated to think one group of people could act so inhumanely to another group. He writes, "I thank God I shall never again visit a slave country. To this day, if I hear a distant scream, it recalls with painful vividness my feelings."

Ironically, the holy books of the three "great religions" are each chock full of references to slavery, and none really consider slavery as a negative. In fact, modern believers will offer explanations to reduce the anti-humanistic message like, "slavery meant something different back then" or that, "the passage was meant to console those in bondage.". According to *1 Timothy 6 1-2*, however: "Slaves, obey your earthly masters with respect and fear and sincerity of heart, just as you would show to Christ. And do this not only to please them while they are watching, but as servants of Christ, doing the will of God from your heart. Serve with good will,…" This quote, which demands both slavery and servitude, was used to justify American slavery up to and through the Civil War.

However intolerant his views on woman's reproductive freedom, Darwin felt a sad kinship to those without means. In *Voyage of the Beagle* Darwin writes, "If the misery of the poor be caused not by the laws of nature, but by our institutions, great is our sin."[285] Social Darwinism is by far the most abusive and ignorant reading of the tenets of natural selection. It assumes a hierarchy within our species based on external characteristics, language, and culture when none exists. It enforces the cultural constructs of institutionalized racism, and at its worse it is a companion to the injustices of slavery and a bane to public policy.

Others would later deeply bastardize Darwin's work to make it fit and justify their racist views. This was something that upset the biologist to no end as he saw the danger of infusing evolutionary biology with ideas of subjective ethnic or racial superiority as applied to freedom from bondage. Darwin was indeed thinking somewhat like a humanist in his approach to the plight of those forced to serve, as well as the poor, even if he did not take much action on those feelings in overt ways.

But turning social Darwinism on its head may have led to the development of some aspects of the social safety net as well as the growth of the welfare state that may have aided the poor and kept poverty at bay. Writing in the *New Statesman,* author John Bew suggests that feelings of national degeneration were occurring in late nineteenth century England. Coupled with advocacy for the poor reaching new levels, the government began social, economic and health programs to help those in need. This in turn lifted people out of abject poverty and raised living standards overall for the poorest of British citizens.

This may be revisionist history, but we cannot discount the idea that some leaders may have chosen to help the poor not out of kindness, but with revolutions and uprising occurring around Europe, more out of the need to maintain what they felt were the cultural standards of British society. Certainly with industrialization there were social and economic reasons as well to help the poor, if for no other reason but to make bodies available for the thousands of factory jobs being created by industrial mass production and new technologies.

Final Thoughts

For some, Charles Darwin could be seen as a Rorschach test for the development of the modernist qualities found within our society today. His observations and work were central up to, through, and way past the publication of *Origin.* He remained a focal point in his own time and remains one even today in our time. Those who laud his work and those who choose to deny it call upon Darwin's name equally. To many, Darwin is a respected secular avatar while to others he is a bane, especially for a select community within our humanity that wishes to push theistic ideas about the world and a metaphysical reality.

Just as future generations will travel a select and different path to their modernity, I imagine that Darwin will be with them just as other great scientists and philosophers are still with us today. For all we know Darwin might even come to see our time and our children's time as both strange and strangely familiar. He would not be wrong, because in fact we owe him so much attribution for current western culture and technological development.

But for all this conjecture one thing is absolutely true. Without Darwin we, and those who surround us would be lesser without this great thinker's work in our lives. Our scientific lives, our philosophical selves, and our humanistic development would be significantly diminished if not for Mr. Darwin and his finches and turtles and slugs and hanging plants. Darwin was the consummate scientist for the nineteenth century. Because of his social status, his means afforded him a lifestyle of gentleman scientist. As both a generalist and obsessively compulsive researcher, it was perhaps his ability to use his time so meticulously which led to the creation of not only new ways of thinking but whole new sciences.

The impact of his timeless writings and intellectual success really does span generations. Though Darwin was a man for his time and not our time, judging him harshly by our standards today would be akin to judging the Romans or Greeks or Babylonians for their "failings" because they did not share our modern ideas or society. Placing Darwin into context does allow us to see him not as a deity, but as a man thirsting for knowledge, a husband and father who knew joy and loss, and a brilliant yet reclusive colleague haunted by the demons of his own fears.

As a man for the nineteenth century and as a man who risked his place in British society for an idea, Darwin must be lauded for his will and his transparency. He is a man who fought his demons as much as he fought the conventional wisdom of his time, when the state and religious authority had the power to end careers, convict a citizen of blasphemy and reject those who they felt were attempting to upend or siphon off their hold on power.

Although Charles Darwin may have been selective in the freethought areas where he placed his energies, we should remember that his five apostles, good men all, were for the most part allied to the Bradlaughs and Besants, the Ingersolls, the Footes, and the Cohens of their day. They eagerly lent support and their names to causes, using their writings and speeches to promote social, economic, and political justice.

Through natural selection, Darwin showed us that nothing stays the same forever—not the land, not society, and certainly not people. Social and biological change is constant. Biological change is unconscious. The outcomes of the two natural elements do combine to bind and separate us in very real ways. Without natural selection there would be no adaptation, no speciation, no next generation. Only extinction. And extinction as Darwin proclaimed does last forever!

In fact, without natural selection all biological life would life grind to a halt. It stops. We stop. We become the end of a line that has persisted on our planet for hundreds of thousands of years. Ask yourself if you want our species to be the species that ends the modernity? Through either our action or inaction our species can create the circumstances in which life on earth may no longer evolve or even exist.

As a species of primate living in the Anthropocene, we own the capability through technology to become the period at the end of the evolutionary sentence. We would then not only diminish ourselves but we would have lost the opportunity Darwin afforded us to view life, nature and each other as fellow travelers on this small oxygen rich floating rock in which we invest our hopes, dreams and plans for the future, a place where we find love, share our joys and our sorrows, indulge our passions, grow and bring new life into this world.

Darwin would pass away at seventy-three in 1882. He worked until the end of his turbulent life. When he finally succumbed to his various illnesses, I believe we can say that he died an old seventy-three. Darwin had been slowing wilting for at least two years, and the culmination of his various pains and illness took their toll. In his lifetime, Darwin did not act as progressively as some of his apostle peers, nor fight openly as did the freethought activists profiled in this chapter. Yet he did remain true to his inquisitive mind and central purpose. That purpose of course was to observe, explore, and explain the natural world in ways that humans before had not been able to understand or conclude.

Fast-forward to the twenty-first century and we see how in our modern lives we are all connected through our devices as well as our common biology. We watch the world on our screens and contemplate others through social media. Our DNA may not have degraded but our attention span has lessened from the thirty-second sound byte to the simple tweet. We must not become disenfranchised from the fights of our freethought brothers and sisters of the past, present or future.

The work remains unfinished. In many places around the globe, women are not free, children are slaves, and there is rampant poverty. The rich, as the saying goes, are getting richer as the poor become poorer. This social inequality is not nature and neither is it Darwinian. It is a humanmade construct where power, status, and stratification inhibit global justice and a truer understanding of our human impact on each other and the environment.

As a social primate we have the ability for injustice and also deeply empathetic good. Just as we have made these human made constructs of social and political power, we have designed tools that allow limitless communication possibilities and democratically based global networks set up to raise the living standards of every citizen on the planet. In Darwin's time and in our own, no matter the front page controversy, there is music to enjoy, museums to visit, art and performances to captivate, and people who earn our respect and awe for the way they present themselves and their ideas.

We need Darwin more today than perhaps at any time in our recent past. Attacks on social justice and scientific modernity, the alt-right idea of "fake news" and "alternative truths," the chants begging us to avoid legal justice in favor of vigilantism have infected our politics like a virus. Rising nativism, anti-globalism and resurgent fascist thinking may be temporarily in vogue in some quarters of the globe and even in some of the most formally progressive nations on the planet. But we are not deterred. For the sake of our common future we must be present for those today but we also must remember those who have fought for our gentler common modernity in the past.

Darwin remains as relevant today as he did 1859, and is firmly aligned to our collective modernity. Like a silent guard, Darwin reminds us that we are temporary, but that the issues of the day are worth fighting for even at much cost. If we are to succeed as a species we better get our act together for the sake of future generations and planet. Otherwise, extinction is an always present danger ready to move our species closer to the exit door of other "also ran" primate species, like the Neanderthals, the Lucys, the Taung children—all gone. Thus allowing for yet another nascent species to, under right conditions and under the mechanics of natural selection, take center stage.

Both the construct of science and the scientists who forge ideas in the kiln of evidenced discovery have a long road and deep burden ahead to bring the Enlightenment to overtly religious communities, as well as to remind secular countries to ensure the flame of reason has enough oxygen to remain well lit.

The burden is not just making scientific information available; it is about making the information accessible. This way someone without much (if any) scientific training could pick up a magazine and begin the journey of understanding, to learn about nature from the point of view of rich intellectual discussion and debate found within the scientific method.

This sense of curiosity is found throughout Charles Darwin's scientific work and writings. Not only are they groundbreaking but they are also easily understood and approachable. While the battle for scientific literacy as it relates to natural selection is played out nationally and internationally in news reports, in the courts, and on college campuses, it is the scientist who maintains the ideas which allow us to learn so much about our cosmos as well as about our own and countless other species.

We are a curious primate species indeed!

CHAPTER 15

THE STORM CLOUDS RISE AGAIN: TWENTIETH AND TWENTY-FIRST CENTURY REACTION TO NATURAL SELECTION

"No apparent, perceived, or claimed evidence in any field, including history and chronology, can be valid if it contradicts the Scriptural record."

Ken Ham, founder, Answers in Genesis

"In science, 'fact' can only mean 'confirmed to such a degree that it would be perverse to withhold provisional assent.' I suppose that apples might start to rise tomorrow, but the possibility does not merit equal time in physics classrooms."

Stephen Jay Gould, Evolutionary Biologist

Introduction——The Times They Are "a Churn'in"

If you like to be entertained while reading let's play a matching game. We will give you a random list of almost forty attempts by anti-evolutionists to interfere with science education, and you try to match the place by drawing a line to the state that has either prosecuted someone for teaching evolution, attempted to stop the teaching of evolution by modifying textbooks, or has entered into or successfully passed legislation to teach intelligent design/creation science next to or as an alternative to biological evolution in public school. **Note: Some states have more attempts than others so you'll no doubt be drawing transecting lines.** (Answers are on page 321).

How did you do? Did you score well? In the end, no matter what you scored the reality is that today the science behind natural selection is being challenged by foes who feel it is their responsibility to defend their religious faith by either ignoring or obfuscating biology, geology, chemistry, climatology and most of the natural sciences in favor of a strict biblical interpretation of reality. They also feel that it is their responsibility

Attempt to Deny, Discredit or Stop the Teaching of Evolution	State/Location
Kitzmiller vs Dover	Georgia
Butler Act	South Dakota
Selman v Cobb County	Pennsylvania
SB561	
Scopes Trial	Tennessee
SB112	Ohio
HB 597	Louisiana
HB368/SB893	Mississippi
Edwards v Aguillard	
1926 Anti-Evolution Act	Tennessee
Assn. of Christian Schools International, et al v Roman Stearns, et al	Wisconsin
Grantsburg Anti-Evolution Policy	U.S. Supreme Court
Epperson v Arkansas	Maryland
McLean v Arkansas	
HB 391	Mississippi
HB 588	Arkansas
Hendren v. Campbell	New Hampshire
SB 6058	Mississippi
HB 4382	California
HB 2554	Arkansas
HB 1286	
HB 268	Georgia
HB 1356	Federal (U.S. Senate)
HB 2548	Indiana
HB 1286	Montana
SB 2692	Louisiana
SB 1386	Oklahoma
HB 1361	South Carolina
HB 2107	West Virginia
HB1531	New Mexico
HB 506	Michigan
HB 679	Washington State
SB 336	Florida
HB 352/SB240	Colorado
HB 106	Texas
HB923	
HB 300	
HB1485	
HB13-1089	

to make you accept these obfuscations, and to teach them to your children. Such ignorance promotes inter-generational nativism, suspicion and the inability to reason when facts are clearly available for knowing.

There was a time in the United States when this nation produced more scientists. Unfortunately, the country now leads the developed world not in mathematics and science education but in incarcerations.[286] [287] [288] America spends more today in real dollars on basic and advanced science research than it did twenty years ago. Yet its citizens have grown more scientifically illiterate. This is because less is spent on science education.

In 1962, the young President John F. Kennedy told Americans, that they would reach the moon in less than a decade. They would meet this noble goal with ships and propulsion systems not invented at the time of the announcement. Sadly, he did not live to see his vision but he was correct. America certainly was, from a science perspective, "the shining city on the hill" that garnered international respect because of the brainpower and will of its government and citizen scientists to reach beyond our world and explore other celestial bodies.

Of course, the initial impetus for all that energy and achievement was Sputnik. In 1957, the Soviet Union successfully launched the first-ever artificial satellite into space, a device called Sputnik. This shocked and appalled the entire United States. From the standpoint of ego, the United States had assumed that the communist Soviet Union could not possibly keep up with the capitalist United States. We assumed that they were behind us in everything. From the standpoint of national security, we were panicked that the hostile Soviet Union would beat us in the conquest of space. We did not want the Soviets to dominate us in control of the stratosphere.

Kennedy's energy, enthusiasm, and financial backing allowed the space program and other scientific ventures to blossom as they never had before. Our post-Sputnik nation embraced the "space race" that put us in direct competition with the Soviet Union for the conquest of space, and specifically for the ability to get to the moon first. Kennedy was cagey enough to offer to work *with* the Soviets in conquering space, but under no circumstances were we going to let them beat us! They didn't.

The near-panic caused by Sputnik led the United States to take science education, scientific research, and scientific development very seriously, for the first time ever. Students from grade school on up were exhorted to go into science. There were scientific careers, and research money available to anyone who could make the grade in a scientific career. It is known that a nation harvests what it cultivates. When the United States deliberately cultivated science and scientists, the nation produced great science and great scientists as a direct result. Can we say with any confidence that it is this way now?

The NASA budget is now about 0.5 percent of the annual U.S. budget.[289] That's down from its high of nearly 4.5% in the mid-to-late 1960's at the height of the space race. [290] Of course we know that that this federal money was to prove America could beat the Russians to the moon, but it was also a time when research and the exploration of space possibilities slid from science fiction to actual national science policy. Where research in medicine is concerned, the budget for the National Institutes of Health has been stagnant since 2003 compared to government revenues, and the budget for the National Science Foundation, which supports non-medical research, has stagnated also stagnated since 2003.[291]

In the more heady days of the 1960s, so much more seemed possible in the public's imagination. Politicians spoke graciously about the wonders of science and bound us to a positive collective future. This binding was not just exclusive to space exploration but it was mixed and meshed with deep interest in other areas of science and medicine, and with interest in our ancient past as well.

While we certainly cannot exclude Raymond Dart's discovery of the Taung Child in the 1920s, and the excavation of *Peking Man* in Asia during this time as well, the 1960s and 1970s proved to be a boon to uncovering our evolutionary past. The Leakey Family and Don Johanson would make their most important hominid fossil discoveries while on excavations in the Hadar Valley in Ethiopia and at Olduvai Gorge in Tanzania.

It was also when perhaps the most notable primatological research would begin as well, thanks to the recruitment efforts of Louis Leakey. He selected Jane Goodall to do her fieldwork with chimpanzees in Tanzania, the late Diane Fossey was selected to research gorillas in Uganda and Birute Galdikas to observe orangutans in Borneo. Each of these brilliant female scientists would help us to understand how close we humans still are to our great ape cousins both psychologically and socially, just as Queen Victoria rightly exclaimed more than a hundred years earlier.

Science makes and keeps us healthy. Science makes us stronger and more able to think rationally about the world which we have inherited. Science gives us the opportunity to make our world even better and more open. Sadly though, in the wrong hands and like many tools, science can be used for destruction.

Advances in our global history have occurred when we have embraced science and not hidden from it. When we unflinchingly supported science and scientists rather than declared that the scientific community has an axe to grind. We progressed when we were politically free to motivate others to think critically and not shrink from our natural curiosity. We moved forward when we invested resources and sponsored creativity for the common good. . However when we are not at our best such suspicions, denials and unfounded accusations against science and the scientific community only serve a narrative necessary to perpetuate ignorance rather than freedom of thought and expression.

For these reasons Darwin and Wallace and their theories are under constant attack, and they need our support now more than ever. Not only in the United States, but also across the planet where oligarchical and theistic governments teach their citizens that evolution is wrong, that it didn't happen, that it is a lie. It is taught at home or abroad that science is a plot by the secular to occupy the minds of religious students, a form of western imperialism bent on destroying the local or national identity of every faith community. People are taught that every true believer should instead put the sword of their faith before the encyclopedia and their religious identity before modernity and must avoid certain sciences areas, like the biological sciences.

So our times they are indeed "a churn'in." The names may change, the scientists working to understand our place in nature and those intent on disparaging science may change, as will the communities and nations that accept or deny aspects of science. What does not change is the ongoing battle to not only control the narrative of science, but science funding, science education, the public acceptance of science and the processes and possibilities which science offers to the world. This must be measured against reason-based ideas which push our knowledge forward, in spite of convoluted theories and theistic alternatives that negate our core strengths which prepare us to work together in understanding the universe and humanity's place in that universe.

It has been more than one hundred and fifty years since the publication of *Origin*. Let's look at how we have responded to Darwin's theory since then. This theory, when denied, establishes a faulty precedent that warps our human understanding of nature . But when it was accepted and understood, it leads to critical new discoveries and fulfills our human ability to remain curious, unbiased, and knowledgeable when it comes to understanding both nature and truth.

Past and Present: Creation Science Evolves into Intelligent Design

A culture war exists between those who cannot accept the theory of natural selection and those who understand its powerful place in nature. This remains a distinctive schism in the social and political lives of citizens of many nations. In the last 150 years, there have been legal attacks on science education related specifically to biology, there have been attempts to teach natural selection and biblical creation as equally valid forms of human development, and there have been millions of dollars spent legislating, many times failing but sometimes successfully passing public science education standards that dismiss evolution or that require biology text books be sold to school districts with front cover stickers stating that non-scientific ideas about creation have equal factual value with evolution by natural selection.

It is not hard to believe that as an academic study, both theology and science are equally worth knowing. However, that does not mean that theology should be taught in science classes, even as an "alternative opinion." Theology can be taught in a history of science class for sure, a philosophy course certainly, in a comparative religion class absolutely. But what if we turned the tables? What if we required every holy book to have a sticker that stated that ideas like natural selection are just as valid as Genesis and should be taught in Sunday school classes? Of course such an example seems like an outrageous violation of people's rights to their faith. Yet, in many faith communities, especially those that are politically and religiously zealous, they see no conflict with imposing their biblical views on public science education within their community and in some cases around the world.

Creation science is at best a banal pseudo-scientific idea which attempts to link science to mythology by trying to prove that biblical creation began around six thousand years ago. This modern religious initiative itself began in the late 1950s and grew stronger along with the evangelical movement in the early 1960s. In 1961, John Whitcomb published *The Genesis Flood,* which led to a resurgence in "creation science" as it called into question the age of the earth. Creation science demands an absolute acceptance of the idea that our planet began only a few thousand years ago, and it rejects the idea of Darwin's theory of common descent.

Natural selection promotes two foundational ideas that are alien to Creation science. First, it shows that there is no material evidence for a divine creator and it further shows that evolution does not require a creator in order to operate. Such biblical heresy is therefore hotly rejected by creationists and intelligent designers. Secondly, it shows that the species that we have today evolved over many millennia. This is the time that chemical change through random mutation and biological change would need for the mechanics of evolution to occur. These mechanics include, but are not limited to, adaptive radiation, speciation, adaptation, and even small and mass extinctions. One or many of these processes happening in tandem with the ecology and geography have produced the many species that we see in modern ecology.

You cannot have a young earth under these circumstances. By its fundamental narrative, creation science also dismisses concepts like the "Big Bang" as being false since it opposes the idea that our cosmos is 14.5 billion years old.

When creation science led to a legal dead end, its promoters invented its newer descendent—intelligent design (ID). It should be understood that intelligent design is nothing more than a strategic re-branding of creationism, for political and legal purposes. ID seeks to promote the old religious argument for a "designer" within all cosmological and biological evolution. In 1984, *Mysteries of Life's Origin* was published by Charles Thaxon.[292] It suggests that the complexity of all life needs a designer not only to get life started, but to make complex modern organisms as well.

This of course is not a new idea. If you recall William Paley's argument for a "Watchmaker" back in 1802, you will see that the argument for intelligent design and its related argument called "irreducible complexity" are just new ways of making the same poor, manipulated and wholly un-evidenced theological argument. Such arguments exist still in books like *Of Pandas and People* (now called *The Design of Life*) and everything written by biochemist Michael Behe.

In 1991, thirty years after the publication of *The Genesis Flood*, professor of law Phillip Johnson published *Darwin on Trial*. Johnson's book has been used by anti-evolutionists for decades and serves as the catalyst for creationist arguments and legislation when seeking to deny Darwin's evolution and replace it with an intelligent design curriculum in science classrooms.

It is important to understand not only the religious arguments for divine creation but also the modern movement which finances and advocates for it. The Discovery Institute, a now global non-profit which seeks to support creationism and intelligent design, is a wolf in sheep's clothing. The institute claims it is an unbiased science and policy-driven think tank, yet it is nothing more than the modern public face and home to biblical "biologists" and theistic researchers supporting intelligent design. The Discovery Institute's other arm exists to socially and legally confront Darwinian evolution. It writes intelligent design bills for local legislatures, advocates for communities that want to teach intelligent design in public science classrooms, and provides "experts" in court cases to advocate for intelligent design when these same communities are challenged by students, parents, and teachers who want to keep the Bible out of their science education.

In 1987, financier John Templeton created The John Templeton Foundation, with its $3.3 billion dollars in assets. His goal had been to ensure a religiously spiritual dimension to science and science research. Since 1987, the foundation has given out almost one billion dollars to scientists, philosophers, theologians, and those working in the social sciences, all willing to link metaphysics to the natural and scientific world. While the foundation calls itself open-minded, in reality it exists to fund an international cadre of religious scientists and scholars willing to take oodles of grant money in order to muddle science and promote the notion of the divine in science education.[293]

Then there is BioLogos, the 2007 creation of Dr. Francis Collins, the director of the Human Genome Project. Dr. Collins is a devout Christian who also serves as the current Director of the National Institutes of Health (NIH). Merging elements of the Discovery Institute and the Templeton Foundation, BioLogos is a squarely Christian organization that has attempted to become a major advocate for science-denying Christians. It unabashedly promotes unscientific thinking by insisting on merging science and faith, specifically Christian theology and Old Testament biblical creation as the supposed basis for all existence. Attempting to merge modern science with theology, their focus remains targeted on outreach and education. They focus especially on children who are home schooled, fundamentalist organizations, and Christian parochial school teachers and their students.[294]

The idea of "teaching the controversy," is a concept put forward by the promoters of intelligent design and creationism. It falsely implies that intelligent design and creationism are equally valid explanations for how species come into existence as evolution by natural selection. Therefore, they say, this "controversy" between the two should be taught in public school science classes. This works to stoke creationist visions for how science should be taught.[295]

The idea of attempting to create harmony between science and faith is also a boondoggle.

Both ideas must be met convincingly through politics, and they should be forcefully and legally challenged when presented as educational legislation. We need science advocates today in much the same way Darwin needed his apostles a century and a half ago. We can also take advantage of the lessons that we have learned from these religious political pressure groups. For instance, instead of religion-friendly state laws regarding public science education, we can make science-friendly laws regarding public science education. For instance, the Massachusetts state legislature is currently considering a science education bill that Abby Hafer put forward. This bill requires that public school science education in Massachusetts must be based on the best available, peer-reviewed, age-appropriate science. This effectively blocks the introduction of pseudoscience, religion, and denialism into the Massachusetts public school science curriculum. We sincerely hope that this bill passes in Massachusetts, and that it then serves as a model for similar legislation elsewhere. We hope that other states follow suit as soon as possible.

The United States—1925 to the Present

The Scopes Monkey Trial

Much has been written about the "The Scopes Monkey Trial," from the reporting at the time in the newspapers, to the legal record, to the play *Inherit the Wind,* to the major film of the same title. The Tennessee evolution trial of the century would in the end crystalize the start of an ongoing modern culture war concerning natural selection and the teaching of evolution.

This culture war attempts to use law and legislation to bend and break the constitutional separation of church and state. It also attempts to separate people into opposing camps that either accept or are repulsed by the theory that humans evolved from older primate lineages without the need for divine creation.

The Evangelical push to legislate away the teaching of evolution has long been one of the main battle plans of religious fundamentalists. Indeed, it was Saint Ignatius of Loyola who in the sixteenth century is credited with saying, "Give me the child for the first seven years and I will give you the man." It is upon this idea that adults can change the course of a child's emotional and educational development by teaching some things as true and others as false, regardless of the evidence, that the whole movement for teaching creationism and intelligent design is founded.

In March of 1925, the Butler Act was passed by the Tennessee legislature. Tennessee state representative John Butler, who was also the Chairperson of the World Christian Fundamentals Association, championed the act. The Butler Act was the first of its kind in the United States and it legally mandated two significant policies concerning public science education. Firstly, it prohibited public school teachers from challenging or denying biblical creation. Secondly, it prevented the teaching of evolution in public school. Teachers found to be teaching evolution could face significant fines and would certainly be in danger of losing their jobs if found guilty of violating the act.

This bill afforded secularists and legal advocates a test case against the growing threat of organized religion in the twentieth century on America's public life and in taxpayer funded education. The American Civil Liberties Union (ACLU) called on educators in Tennessee to challenge the act and offered to finance any legal action against it. Enter John Scopes, a twenty-five year old high school substitute biology teacher who on May 5, little more than two months after the Butler Act was passed, was charged with teaching evolution to his students in Dayton, Tennessee. Scopes' bail was set at $500.00 and was paid immediately by Paul Patterson, the owner of the *Baltimore Sun* newspaper.

For the eventual prosecution, the state called on several minor, midlevel, and then major attorneys of depth and breadth. First, local attorneys Herbert and Sue Hicks were given the case. Two more experienced lawyers would eventually handle the prosecution; attorney Tom Stewart would lead it. Stewart was certainly an anti-evolutionist as he believed that evolution was "detrimental to our morality" and served as a direct assault on Christianity.[296]

The other attorney was Williams Jennings Bryan, who ran for the presidency three times, lost each time, and was the former U.S. Secretary of State under Woodrow Wilson. In 1925, Bryan would be sixty-five years old and would serve as senior advisor and chief counsel to the prosecution. Bryan was a long-time anti-evolutionist. As early as 1905 he stated "Darwin's theory represented man working coldly on the operation of hate" and insisted that Christian love and the social gospel was man's only hope for redemption.[297]

For the defense, the ACLU and Scopes turned to famed attorney and skeptic Clarence Darrow. Darrow's pedigree made him a natural foe of the anti-evolutionists. His family served in the American Revolution and his father was an anti-slavery freethinker. His mother was involved in the suffrage movement. Darrow fought for labor rights but eventually left that area of practice when it was revealed that he was allegedly involved in a plot to bribe a juror. Darrow was however a successful defense attorney who, in 1924, got Leopold and Loeb, two teens convicted of murder, life sentences of life in prison rather than the death penalty.

But Darrow was not alone at the defense table. Darrow was joined by some very experienced and polished attorneys as well. They including Arthur Hays, who helped found the ACLU, as well as Dudley Malone. Malone was a successful divorce attorney who maintained a very strong interest in civil rights throughout his legal career.

The judge in the Scopes trial was local magistrate John Raulston. Judge Raulston was seen as someone who favored the prosecution throughout the case. He often butted heads with Darrow because of Darrow's antagonism and frequent off-the-cuff remarks about the prosecution's arguments and how the course of the trial was being handled. This perceived favoritism of anti-evolutionists was not new. Even the Huxley-Wilberforce debate of 1860, while not a legal proceeding, was supposed to be a scientific one. Yet Henslow allowed the proceedings to take on the atmosphere of a circus, allowed an ill-informed priest to carry on at length at a scientific meeting, and it did place Darwin's theory on trial in an academic sense. This fear concerning the impartiality of judges would show itself time and again in later legal actions concerning the teaching of evolution.

The defense of Scopes went through several phases and alternative narratives. The ACLU originally presented the case on constitutional grounds, arguing that Scopes's First Amendment rights as well as his academic freedom were imperiled by the Butler Act. They also tried to expunge the charges against Scopes by claiming there was little difference between biblical creation and science by using the "God started it all and science is uncovering it" defense. But none of these defenses were gaining sympathy from the judge.

The defense led by Darrow then employed a new method in their attempt to dismantle the prosecution's arguments. Darrow's witnesses were in fact never put in front of the jury as the seated jury was not allowed in the courtroom during the reading of the defense's testimony. It was at this time that Darrow asked his most unusual request: would Bryan serve to be vetted by the defense because of his advocacy, expertise, and admiration of the Bible?

Now with a packed court, upon Bryan's acceptance Darrow mercilessly asked questions about the literal nature and reliability of the Bible. Was the story of Adam's rib a metaphor or real? Was Eve's acceptance of the snake's apple true? Where did Cain find a wife? Many nuances of biblical stories were explored that fell well outside of historical fact and science's ability to judge physics rather than metaphysics. These blistering attacks on the credulity of the Bible exhausted Bryan.

The defense questioning was daunting and it was clear that even for a Bible "scholar" that the nuances of religious belief had no place in the science classroom, especially if it was based on personal faith and the specific biblical interpretation of one religion. Judge Raulston concluded that Darrow's examination had little to do with Scopes's guilt or innocence so the testimony was struck from the case. But that also meant that the prosecution could not cross-examine Bryan and perhaps mount a counter-attack.

In his summation, Darrow spoke to the jury and explained that the Judge had essentially bound the defense's hands by narrowly limiting the evidence that would exonerate Scopes. Nothing was offered to the jury about evolution's accuracy and truth. From the beginning of the trial to its end the Judge only allowed testimony that answered the simple charge, did Scopes violate the Butler Act.

For the prosecution, Bryan made the claim that evolution nullified civilization and also precluded goodness, as he told the jury,

> If civilization is to be saved from the wreckage threatened by intelligence not consecrated by love, it must be saved by the moral code of the meek and lowly Nazarene. His teachings, and His teachings alone, can solve the problems that vex the heart and perplex the world.[298]

No one doubted that Scopes violated the law, so the only answer the jury could find was guilty. It took them less than ten minutes to deliver the guilty verdict. The answer as to whether the act was fair or constitutional was not allowed in as evidence for the jury to consider or include in their deliberations.

John Scopes was convicted, fined, and ordered to pay $100 for violating the Butler Act. However, Scopes would receive some justice as the case was immediately appealed to the Tennessee Supreme Court. While the court did not weigh in on Scopes's innocence or guilt, the guilty verdict was overturned on appeal because of a technicality. The court found that the jury rather than Judge Raulston should have set Scopes's fine. A fine that could not under Tennessee law have been more than $50.00, if set by any local judge.

The Evolution Debate—-A Forty-Year Cooling Off Phase

While other states passed similar laws to Butler Act, other national and global emergencies would take center stage. The Great Depression would require America to rework its financial, social welfare and educational systems to avoid total economic catastrophe. Coming out of the Great Depression, America and its allies were then forced to confront the evil nature of fascism across Europe.

The United States would delve deeply into its national will to fight a second war in Europe and a simultaneously devastating war in Asia, all to ensure secular global peace and security could be won against the forces of Nazi and Japanese threats facing constitutional democracies.

Perhaps America and its educators had optimistically thought (if they thought about it at all) that with the Scopes trial now firmly in the past and the turn towards science and modernity after World War II meant that public and university education would be based on hard science and evidence. The United States next battled the Soviets and Sputnik by pouring millions of dollars not only into science and science education at the K-12 level, but also millions at universities to enhance the nation's brain trust when it came to science and scientific discovery.

With the coordinated political rise of the Evangelical movement in the United States in the 1950s more and diverse religious challenges began to take center stage as the sophisticated legal arguments for teaching creationism and intelligent design evolved. At the same time, brave educators as well as long-established civil rights organizations would begin to challenge local school boards, larger communities, and even state laws that denied students the right to know and learn about natural selection and Charles Darwin.

The Beat Goes On: Things Begin to Change in the 1960's

Epperson v Arkansas

The decade of the 1920s was a time in the Deep South when religious forces not only led their communities from the pulpit within houses of worship but they also led them within their legislatures as well. So while Tennessee was the first state in the nation in 1925 to propose and pass anti-evolution teaching bills it was not the last. Mississippi made teaching evolution a crime in 1926 and Arkansas did so as well in 1928.

Some four decades later in 1966 a high school biology teacher, twenty-four year old Susan Epperson, working in Little Rock, found herself in violation of Arkansas's 1928 anti-evolution law. Because of changes in both science curriculum and their newly recommended science textbook, Epperson and her colleagues included information on both Charles Darwin and natural selection in their teaching.

This was in violation of the Arkansas statute that made it illegal to teach evolution in public schools. Therefore, controversy ensued.

This time however, Epperson's criminal defense did not rely on just imaginative and charismatic orators to defend the allegations of criminality for her actions by teaching evolution. The National Education Association (NEA) as well as the local chapter of the American Civil Liberties Union also backed Epperson legally. The legal filings asked the court to extinguish the 1928 law as unconstitutional as well as adding an injunction to ensure that she could keep her job.

The Chancery Court found for Epperson on the grounds that the 1928 law violated the First and Fourteenth Amendments of the U.S. Constitution. However, the state appealed the ruling of the lower court to the Arkansas Supreme Court whose wisdom it was to reverse the Chancery Court's legal decision. The Supreme Court did not invalidate the 1928 law because it had an opinion on evolution per se, but because it held that the state has the right to set its own curriculum standards and by disavowing the law, it meant that the state could no longer control its public education curriculum. Essentially, they saw this as a state's rights issue rather than a national or local issue.

Epperson and her legal team then appealed the Arkansas Supreme Court decision to the United States Supreme Court. Presiding Justice Earl Warren and the Warren court heard the case and upon review reversed the Arkansas high court. The Supreme Court found that the 1928 statute violated Epperson's and other teachers' free speech rights and that the anti-evolution prohibitions in the law existed to protect very specific religious points of view, thus holding that such a law violated the Constitution's separation of church and state.

Talk'in About my Generation—Late Twentieth and Early Twenty-first Century Challenges to Teaching Evolution

Edwards v Aguillard

To assume that the creeping nature of non-science and a creationist worldview is something that belongs to history would be wrong and intellectually dishonest. It would negate how at the close of the twentieth and the start of the twenty-first centuries, religious forces are still working to dismiss Darwin and evolution in favor of theology-based science, attempting as always to gain a legal foothold for teaching creationism and intelligent design in public schools and within public science standards in education.

In 1987, the Supreme Court was again visited by the creation versus evolution debate. Now it was Chief Justice William Rehnquist's court that would find yet another state violated the First Amendment's Establishment Clause. In 1981, the Louisiana legislature passed the *Balanced Treatment for Creation-Science and Evolution-Science Act.* As the bill suggests, it required that public school science classes teach both biblical creation and Darwinian evolution as completely equal and factually validated theories of how life on earth and the universe were created.

This theistic bill was developed by the Discovery Institute and was intended not to teach biblical creationism per se, but offer divine creation as an alternative choice to students as they studied biological evolution. The crafters of the bill along with the author, State Senator Bill Keith, developed language so that it read like an unbiased piece of legislation which was to foster academic freedom rather than its actual purpose, which was to equate Old Testament theology with verifiable and observable science.

Enter yet another brave public servant, Don Aguillard, who was a high school science teacher for many years and who specialized in instructing students in advanced biology topics. Aguillard found that having to teach creation science along with evolution actually limited the intellectual freedom of students, in part because the act insisted in offering only one specific story for non-scientific creation. In a lawsuit brought against the state to invalidate the act, the Fifth Circuit found the Creation Act violated the First Amendment's Establishment Clause and was struck down.

Undaunted, Louisiana's creationist lobby believed that they could win on appeal and petitioned the Supreme Court to hear their case. In a seven to two decision, the Supreme Court found the Louisiana law was unconstitutional and agreed with the Fifth Circuit that the legislation had to be struck down.

The court found that the law did not, as its drafters had insisted, support academic freedom, but simply called for the teaching of specific theological precepts in the public realm, specifically in science classrooms. As with the lower court decision, the Supreme Court found the law violated the Establishment Clause of the Constitution.

However, one of the two dissenting judges, the late Justice Antonin Scalia, who was an ardent Catholic, creationist and believer in angels and demons, found the law acceptable and was friendly to the Louisiana creationists.

The Santorum Amendment

Another ardent Catholic, politician and former Senator Rick Santorum (R-Pa), attempted to infuse and expand creation science and intelligent design curriculum throughout the United States in one fell swoop. In 2001, while then President George W. Bush was remaking federal education policy with his "No Child Left Behind" legislation, Santorum and the Discovery Institute crafted an amendment to the bill as it went before both the House and Senate.

The aim of the amendment was to again use a wedge strategy to boldly (and falsely) claim that evolution in general and curriculums that teach natural selection in particular remain controversial topics in science. Each, they insinuated, was not widely accepted. In fact, the only place where evolution and natural selection are controversial and not widely accepted are in the minds of creationists, proponents of intelligent design, the creationist lobby and organizations and their legal representatives.

Fortunately for America two things happened which nullified the amendment. The first was almost one hundred educational and scientific organizations lobbied Congress to reject the additional legislation, which they did stating,

The conferees recognize that a quality science education should prepare students to distinguish the data and testable theories of science from religious or philosophical claims that are made in the name of science.[299]

The second was that Sen, Santorum was not reelected to Congress and all his main congressional supporters of the amendment, including former Speaker of the House John Boehner, also are no longer serving in Congress as well.

But federal legislators haven't given up the fight to "teach the controversy" and it appears that governors and senators (all Republicans) still see intelligent design as their way to Christianize and shape the minds of young people using public monies to teach creationism in public schools. In fact, the former Republican Governor of Indiana and current Vice President, Mike Pence, has stated the following: "Only the theory of intelligent design provides even a remotely rational explanation of the known universe.[300] So it is clear that almost a hundred years after the Scopes Trial, elected government officials in some areas of American politics—in local, state, and national venues—haven't given up but instead have doubled down on creationist ideology.

As Donald Trump has now taken office and will lead the United States for the next four (possibly eight) years, his cabinet selections, especially for the Department of Education, are telling as to where the government may move with regards to federalizing intelligent design ("ID") as a companion or outright alternative to teaching evolution in public school settings. Certainly, the selection of controversial conservative billionaire Betsy DeVos, an advocate for charter schools, whose husband Richard is a devout ID proponent,[301] is somewhat chilling. Although there is no crystal ball or smoking gun the likelihood the intelligent design will gain federal credibility is not an unreasonable assumption to make based on the statements of the DeVos family.

A federal nod in support of the Evangelical cause occurred even as we worked to complete this book. When Reverend Billy Graham died in February of 2018, Congressional leadership requested and President Trump obliged to have his body lay in state in the Capitol Building. Such an honor for a man who used the Bible and his faith to call whole swaths of people sinners, who denied science and evolution, and who was ardently against civil rights, seems to be an affront to not only civility, but to all those who laid in state previously and did so deservingly.

The Dover Decision— The Fight Gets Personal Pitting Neighbor Against Neighbor

One of the newest and boldest strategies in the "teach the controversy" arsenal of the Discovery Institute is to work on the local level to infuse intelligent design curriculum in public education. The focus has shifted from crafting local, state, and federal creationist legislation to micro-targeting school boards that may be open to intelligent design curriculums. The goal is to find these boards if they are already constituted or to work within the community to get new school board members elected who are friendly to ID. This way, a board member (or members) can introduce new "science and academic freedom standards" that include intelligent design curriculums to be used in classrooms replacing evolution or requiring ID pseudo-science be taught alongside evolution as part of their local public school's science education.

In 2004, this is exactly what happened in Dover, Pennsylvania, as the school board chose to include the creationist textbook, *Of Pandas and*

People, in their science curriculum and wrote a press release which stated that at the start of the school year high school science teachers would be required to read a prepared statement in their classrooms stating that evolution was just a theory, and that intelligent design is a possible alternative explanation of life which differs from Darwin's view.

The board's statement implied that natural selection was more of a guess about how animals and humans evolved rather than a scientifically based theory supported by observable science and bolstered by decades of research, impartial geological findings, the fossil evidence, and advances in chemistry and biology which all point to the correctness of descent through modification. In their way, the school board was essentially doing what creationist legislators and the Discovery Institute couldn't do—import intelligent design directly into secular taxpayer funded education curriculum by fiat.

The school board vote for ID was not unanimous. Three of the board members resigned after the vote, choosing to not support the new initiative during their incumbencies. In less than a month after the board's decision, the American Civil Liberties Union along with eleven parents, including lead litigant Tammy Kitzmiller, filed suit against the Dover school board to stop the new policy from entering into effect on the grounds that teaching intelligent design violated the Establishment Clause of the Constitution.

The coming legal battle mirrored the Scopes Monkey Trial in many ways. It garnered a great deal of national attention. It again placed Darwin's theories in the forefront of the American consciousness. Finally, like Scopes, the Dover trial had its theatrics and dramas playing out on small and large stages within the community. Many families felt alienated from one another because of the stance they took for or against ID, and many friendships were tested and harmed before, during and even after the trial.

To be sure, the new ID rules also ran counter to Pennsylvania's science standards that taught the concepts of Darwinian evolution on state exams in order for high school seniors to graduate. ID frightened many Dover parents not just because it was a foreign concept but also because it remains scientifically un-evidenced. They saw ID as dumbing down the Dover curriculum and in effect dumbing their children down as well. The parents also feared that the ID curriculum would make their children less competitive within the state and less prepared for college. In addition, many community members did not want Dover to be considered a laughingstock in Pennsylvania or the nation, but the ID curriculum and the subsequent trial was indeed making that happen.

For those parents and community members who did support the school board, they saw their opponents in lawsuit as a group of intellectual ne'er-do-well parents and legal outsiders. They felt that the litigation interfered with their community's right to teach their children alternatives to evolution that they felt were important to know. However, these folks were certainly not in the majority and most were Christians who saw ID as affirming their faith through science.

The witnesses for both the plaintiffs and the defendants were rich with names from both sides of the intelligent design debate. In just over six weeks, twenty-three witnesses were deposed for the plaintiffs. They included all the parents named in the lawsuit, molecular biologist and Brown University professor Kenneth Miller, the Catholic theologian John Haught, and author Barbara Forrest, whose book about the Discovery Institute detailed its plans regarding intelligent design and the wedge strategy it developed to "teach the controversy."

For the defense, Lehigh University biochemist and creation scientist Michael Behe spoke about ID and his concept of irreducible complexity. He admitted under cross-examination that his concept and ID and irreducible complexity had received no peer review and thus offered little collaborative scientific evidence to his argument. In addition, when philosopher Steven Fuller spoke and attempted to equate ID with the civil rights movement, he was viewed almost as a zealot and his testimony failed to win many proponents.

A month before the end of the trial and in a rebuke of the ID policy, none of the Dover Area School Board members who had voted for the intelligent design curriculum were re-elected. Clearly the plaintiffs had spoken well at trial. And in November 2005, so did the community with their right to vote for new school board leadership.

In December 2005, Judge John E. Jones III (an appointee to the federal court by President George W. Bush), after hearing all the witnesses, handed down his legal decision to a packed courtroom. Just prior to and during the trial, the plaintiffs and their legal counsel feared that Judge Jones would not be unbiased. Perhaps he was already in the pocket of religious conservatives since he was appointed by a born-again president who openly stated that both ID and science should be taught equally. However, those fears were unfounded as Judge Jones dealt the Discovery Institute and ID proponents a stunning and irrefutable legal loss with his decision.

In his 139-page ruling, Judge Jones wrote very directly and very plainly. He found that intelligent design was not based on science and that in fact it was creationism and theology. He also ruled that the *Pandas and People* textbook made false claims and used untruthful examples, that ID essentially was a religious argument for the basis of life, and that ID had been successfully refuted by the scientific community.[302]

Judge Jones also found that the School Board's actual purpose was not a secular intent to foster academic freedom but a religious attempt to place theology in public school classrooms. He also noted that by creating the ID policy, the school board had failed in its elected capacity to support the Dover community.

Each of these rebukes—of the Board and its ID policy—of the Discovery Institute and its ID claims—of the witnesses for the defense—of the *Pandas* book—helped form the final legal conclusion that intelligent design is theology and that the implementation of ID curriculum in public school classrooms violates the Establishment Clause of the United States Constitution.

More Anti-Science Legislation Since the Dover Decision

In 2008, the Louisiana legislature passed the infamous Louisiana Science Education Act. This allows public school science teachers to teach creationism and intelligent design and other non-science using "supplemental materials"—which will, of course be generously provided by the Discovery Institute. This law is on the books to this day.

Likewise, in 2012, Tennessee passed a law that allows public school science teachers to teach creationism, intelligent Design, and climate-change denial (as opposed to any other topics!) as challenges that will promote "critical thinking." It should be noted that this critical thinking is only supposed be applied to topics that the legislators don't like, as opposed to being an honest and broadly-based critical thinking curriculum. This law is also currently on the books.

Other states have also made a point of trying to pass legislation of this type, with the kind and generous help of the nice folks at the Discovery Institute, who provide both model legislation and legal help with these efforts.

Charter Schools, and the U.S. Home School and Good News Club Movements

Working against well-funded theistic organizations dedicated to supplanting science with creationist religious ideology is a never-ending battle. It's sort of like the carnival game "whack-a-mole." No matter how many times you hit one of those rodents on the head forcing him back underground, another one seems to pop up somewhere else with ease.

There are three growing areas of national and international concern as the intelligent design movement yet again evolves and adapts (ironic to use these terms). The troika includes the charter schools movement, the increase of home schooling, and the after-hours use of public school facilities for so-called Christian Good News Clubs. Each uses the public's goodwill, and in some cases even worse, public tax dollars to fund creationist views and to spread biblical versions of the creation of the universe, the earth, and our very humanity. This is done outside the confines of biology class, but in social and educational settings where law and parental involvement is lacking.

Charter Schools and Private Schools Using Public Voucher Programs

The expansion of the charter school movement comes from a growing political response to real and imagined academic deficiencies and the public's mistrust of public education. Since the start of the new millennium, charter school enrollments have grown from about four hundred thousand pupils to almost three million students in 2016. Not all charter schools are equal in academic quality, some are better and some worse than the public schools which they are supposed to replace. In many cases they are for-profit and it must be noted that not all of them teach creationism. However, the one thing that they all have in common is their funding source, and that is your local, state, and federal tax dollars.

Every charter school must apply and be registered within their state to provide academic services to primarily underserved communities in which they work. They also agree to support state educational standards in order to receive their operational funding to serve the public's interest. However, oversight of both curriculum and management of the schools is weak across all levels of government. So much so that in the 2014 *Annenberg Institute for School Reform* report, the investigators found the entire charter school system rife with failures.[303] These include failures in maintaining curriculum and governance standards, failures in executive transparency and daily management, failures in accountability and equity, failures in protecting student data and failures in the enrollment and retention of students.

For those who intend to teach ID in their charter school's science curriculum, and most can be found in the southern United States, when governmental oversight is weak or when they serve in communities where creationism is championed, those who wish to game the system or obfuscate their actual intentions can do so almost unimpeded and without much challenge.

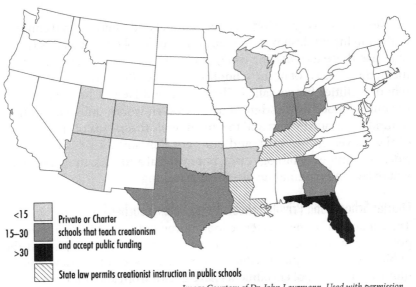

<15   [ ]   Private or Charter
15–30 [ ]   schools that teach creationism
>30   [ ]   and accept public funding

[ ]   State law permits creationist instruction in public schools

Image Courtesy of Dr. John Laurmann. Used with permission.

However, there are many more private schools, most of them parochial in nature, which take public money through voucher and scholarship programs. These programs allow poor or underserved students to gain what is perceived as a better quality non-public education. But if charter schools receive little governmental oversight, one can imagine that private schools receive even less intrusion in their curriculum and overall management. This presents an ongoing problem and as such allows for the active non-teaching of science by creationist proponents unheeded by normal checks and balances, oversight, audits or accountability found in public education.

The above image represents the number of states that allow public, charter, and private schools to teach ID and receive public funds.[304]

Home Schooling——Parents Teaching Biblical Creation as Both History and Science

Many parents who decide to teach their children at home rather than send them to public school are deeply suspicious of public education and secularism in general. They may claim that their local schools are poorly funded, are mismanaged, or have insufficient curriculums, and to be sure in some cases this may be true. However, when you combine a creeping suspicion of government with the simultaneous need to teach one's faith as the sole truth in regards to both history and science, one eliminates academic standards and eviscerates modernity in favor of religious dogma for the sanctity of an evangelical worldview.

According to data published in 2015 by the U.S. Department of Education, the number of parents that homeschool their children rose more than sixty percent in the decade from 2003–2012. That totals almost 1.8 million children who are being taught outside of public education systems. The National Center for Educational Statistics (NCES) also surveyed parents as to why they chose to teach their children at home rather than send them to public school. The three most popular answers included, "Concern of the School's Environment" (34%), a "Dissatisfaction with curriculum and instruction" (17%) and "Desire to provide religious instruction" (16%).[305]

With almost two million children available to be misinformed about evolution, there is a huge market for the sale of creationist textbooks, intelligent design curriculums, as well as numerous evangelical parenting support groups offering advice on how to infuse faith with both science and history. These print and online educational materials as well as the intervention services are especially useful for those creationists who teach their children at home and who accept the concept of a young earth while disavowing natural selection as a trick by both secular scientists and Satan.

The Good News Clubs—Not Really Such Good News After All

Good News Clubs are the missionary arm of the modern evangelical movement. Their purpose however is not to visit some far off land and preach the gospel. It is to come into urban neighborhoods and offer recreational services to minority and underserved populations in cities across America.

The goal is not to provide secular solace but to subtly infuse their theological views through smiles and games focusing on populations most vulnerable because of social economic disparities in income, access to education and safe leisure spaces. When and where do Good News programs usually make their services available? Primarily it is after 3:00 pm, as after-school programs inside the same schools from which the children were just dismissed.[306]

Evangelicals who manage Good News programs know that there is power and authority within the school setting for children and their parents alike. So by offering services in the school buildings they graft onto the school's stature. Parents frequently assume the program has the school's endorsement as being both fair and equitable. Because providers are not usually directly preaching while servicing children after regular school hours, there is little action that can be taken in terms of church and state separation simply because regular school isn't in session.

After regular school hours, the school building reverts from a place where secular education is mandated to a "facility" open to all community groups wishing to rent space. For the Good News Club programs, it's not necessarily about immediate biblical indoctrination so much as it is to be seen as a place of emotional and physical safety for child and parent alike. The building of such trust then allows for the program to have a long-term relationship with pupil and parent, and then indoctrinate them in evangelical positions and biblical scripture.

Good News Clubs present themselves as an oasis in the urban desert. Since children are not required to attend, do not have to stay in the program and are free to leave at any time, complaints about the evangelical nature of the clubs is somewhat muted. But children and their parents really face a Hobson's choice regarding attendance if not full participation. There are few alternative safe spaces for urban latchkey children whose parents or caregivers are working full-time, sometimes holding two or more jobs just to survive.

It's not just creationism that the Good News Clubs push. They also are emphatically against gay marriage. They ensure that their literal views on scripture and the Bible are at times the only literature accessible to students while in their programs. They subtly teach young and impressionable students to be suspicious of any authority, parent, teacher, or the government, if they disagree with Good News publications and instruction.

International

We have seen many instances where Americans of religious faith attempt to discredit Charles Darwin and the theory of natural selection. However, America is not by any means the only nation to face continuous denial of evolution from its citizens. In the United States, intelligent design fails to gain a foothold in public educational curriculum due to the constitutional separation of church and state found in the First Amendment of the Constitution.

Because the United States was established as a secular democracy without any formal state religion it cannot directly provide tax monies or favor governmental policies to support any particular faith tradition. It also cannot favor one religious tradition over another. Such action would also immediately violate the Establishment Clause of the Constitution.

This is not true in many other nation-states, western or otherwise, governed in full or in part by secular democracy or forms of oligarchy, dictatorship or even via full-on theocracy. Many modern governments provide public money for religious education in public schools.

These include nations with governments which have long-held ties to one or several faith traditions. England, Romania, Turkey, Canada, Poland, Greece, Germany, Finland, Austria, Thailand, Burma, Pakistan, India, Russia, and several of the Baltic states all provide public monies to public and private schools in order to indoctrinate children into a particular faith tradition.

Many of these national governments with centralized education policies also offer their students (and parents) the option to take ethics classes instead of religion classes if they so choose. But the social stigma of opting out can be daunting and most children just want to be with their friends regardless of the school venue. Sadly, and in too many instances, if a child cannot fit an alternative ethics or philosophy class into their schedule to replace the religion course they are often faced with just wandering the halls of their school until their next class begins.

But creationism is alive and well in these and other nations. Most if not all Muslim nations demand adherence to the teaching of the Qur'an that firmly disavows natural selection.[307] In fact, evolution is seen it as a form of biblical heresy and western imperialism. Both the Sudan and Saudi Arabia ban teaching evolution outright. Such heresy is punishable and can include fines, loss of job, imprisonment, or even physical harm. Even just the thought that a person could face such terrible retribution for accepting evolution is usually enough to cease any fair exchange on the topic at home, in government and in schools and universities. Even the more moderate Turkey teaches creationism and intelligent design as science.

In a 2011 survey commissioned by Reuters, their poll found about 12 percent of Germans self-identify as creationists. In fact, one of Germany's top education ministers, Karin Wolff, suggested strongly that intelligent design should be taught in biology class. Similar beliefs concerning science and religion have caused other top European education officials to the replace standard biology textbooks with books that teach intelligent design (this occurred in Romania, Switzerland, and Russia). In Asia, both South Korea and Australia each have regions of their nation where biology textbooks have been exchanged for creationist ones.

Like all forms of ignorance, the open denial of evolution is multinational, multicultural, and multifaceted. But ignorance cannot fill the gaps of curiosity, wonder, and practical economic stability that come from our understanding of science and the demands placed on us all in our modern world. So it should come as no surprise that in America those states that deny evolution are amongst the poorest and least healthy, and internationally, the nations which seek to deny evolution are those without secular democracy, have weaker economies and GDP's, are their citizens are less healthy and poorly educated.

The year 2017 was a defining year for how we choose to recognize and teach evolution and natural selection. In January 2017 the nation of Turkey, a formally secular nation that has prided itself as bridging Europe and the Middle East in terms of attitudes, religion, culture, and thought, tilted towards religious hegemony. What is the result? New public high school textbooks have been reviewed and edited. Evolution and the concept of natural selection have been removed from the teaching standards and requirements across all public school curriculums in the nation.[308] This omission breeds ignorance and signals a national institutional shift towards embracing creationism and away from the reality of natural selection.

Conclusions

The diversity of scientific ideas and religious criticisms concerning evolution remain almost the same as it was in 1859. Even though the times, names, and places have changed, the embracing or rejection of descent through modification exists as strongly today as it did in the nineteenth century, outside of scientific communities. At first Darwin's theory was introduced to America and Europe, where it was discussed and debated by laypersons, intellectuals, scientists, and theologians. But as modernity and information sharing has expanded across the globe, the theory of evolution has continued to bring religious ideas concerning metaphysical creation into direct conflict with teaching evolution and with the scientific community around the world.

The issues of how Charles Darwin and natural selection are perceived, as well as how evolution is understood and taught is more than a scientific issue. Because if it were just a scientific issue the criticisms and arguments would be based on claims made by those who know and use the methods of science best, fellow scientists.

Because the theory of evolution has a direct impact on science as well as on theistic ideas of how humans and the universe developed, for the last one hundred and sixty years the "debate" about natural selection has also been driven by ego, politics and subjective community standards. When trying to belittle Darwin's genius most of these negative efforts have attempted eliminate or coopt evolution or equate it with other non-scientific ideas such as intelligent design.

Interestingly, The Interacademy Panel (IAP), which serves as an international consortium of scientific academies, put out a lengthy statement about teaching evolution. It was signed by sixty-eight leading national scientific academies from around the globe and it declared,

> We, the undersigned Academies of Sciences, have learned that in various parts of the world, within science courses taught in certain public systems of education, scientific evidence, data, and testable theories about the origins and evolution of life on Earth are being concealed, denied, or confused with theories not testable by science. We urge decision makers, teachers, and parents to educate all children about the methods and discoveries of science and to foster an understanding of the science of nature. Knowledge of the natural world in which they live empowers people to meet human needs and protect the planet.

> We agree that the following evidence-based facts about the origins and evolution of the Earth and of life on this planet have been established by numerous observations and independently derived experimental results from a multitude of scientific disciplines. Even if there are still many open questions about the precise details of evolutionary change, scientific evidence has never contradicted these results.[309]

What this suggests is that there remains a huge disconnect between scientists, science educators and the general public. Indeed, in surveys we find that while 98 percent of scientists accept biological evolution as real and factual, while in the United States it is only about 60 percent accepted by the general public, and a full third of that number believe evolution was divinely inspired.

Polls suggest that belief in God and miracles is declining steadily, yet the funding and efforts to enhance both student and public science education seem maudlin at best. In addition, nations around the globe vary greatly from almost full acceptance of evolution to almost full denial of the Darwin's theory in efforts to preserve a biblical view of creation. Such inconsistencies show that acknowledging natural selection and teaching biological evolution is a socially complex set of issues that demand our time, attention, and advocacy.

It is only through consistent confrontation of the ignorant and standing guard against the willful and purposeful misinterpretation of Darwin's entire body of work that we can hope to turn the page and fully contemplate what Darwin and his apostles have given us in terms of growing our knowledge and understanding. Not just of ourselves, that would be egotistical, but all of nature. Combined with today's modern science, we can expand on the fundamental and essentially rudimentary insights offered by Darwin. We can through science build even greater knowledge of our humanity, our place on our planet and within the cosmos.

Science does not willfully dethrone any particular religion but it does help us place the need for faith into social and psychological context. However, science does allow us to free our individual and collective minds to explore larger and greater ideas about the natural world that will substantially bring us greater insights, and perhaps even a more benign global civilization.

Because once we accept our material place in nature we can see ourselves as equals, forgoing national boundaries, phony racial categories and subjective ethnic and cultural differences which often serve to divide rather than enjoin us in a common humanity. If we can get past all of that baggage, then science and science education are the best tools we have invented to build for the future using reason to create a better understanding of who we are and where we may be going.

CHAPTER 16

IN CONCLUSION: DARWIN, THE SCIENCE OF EVOLUTION, AND THE WIDER WORLD

Science and the Millennium

On January 1, 1999, the British Broadcasting Corporation (BBC) named William Shakespeare as its listeners' choice for their Person of the Millennium. We would argue that Galileo and Darwin were far more important. While Shakespeare described our world in sonorous language, Galileo and Darwin gave us a completely new way of *seeing* it. This new way was vastly more productive.

These contributions to human life are in no way equal. If Shakespeare had never lived, theater would still have existed. But without people bravely setting the scientific revolution in motion, modern science would not exist. The scope and effect of the scientific revolution cannot be underestimated. In the four hundred years since it began, we have learned far more truth about the world than we did in all of our preceding history, and science has changed human life beyond recognition, infinitely for the better.

The early scientists proved the primary importance of looking at the world carefully, with your own eyes, and reporting what you see, not what you are expected to see. This distinction has rarely been as important as it is today, when our own president tells us to believe what he says, rather than believe the evidence provided by our own eyes. All our instruments and all our careful experimental methods and statistical analyses are simply ever more powerful ways of seeing things accurately.

Darwin

The scientific revolution begun by Galileo and continued by Newton was brought to full flower by Darwin. In one masterful work, *On the Origin of Species by Means of Natural Selection*, Darwin provided a lens through which the entire biological realm could be viewed and comprehended. Suddenly, the entire biological world made sense. Everything from the tiniest protozoans through large, complex ecosystems that span many hundreds of miles could all be understood as being products of evolution. With the idea of common descent, classification suddenly made sense in a grand, overarching way, which it had not done before. When Mendelian inheritance was re-discovered in the early 1900s, it, too, made sense in the light of evolution.

Even the field of physics was corrected and pointed in the right direction by understanding evolution. Lord Kelvin, the preeminent mathematical physicist during Darwin's time, calculated the age of the sun and the age of the earth. He did it wrong, and calculated ages that that were far too short for evolution to have taken place. In the end, Kelvin's tiny age estimates were shown to be wildly wrong. Meanwhile, the evidence for evolution has gone from strength to strength.

Evolution in the Twenty-first Century

Since Darwin's time, evolution has continued to clarify biology. Fields that were not even imagined in Darwin's time have shown the truth of Darwin's science. Molecular biology had not yet been invented, but the findings in that field continue to show us that Darwin was right.

For instance, many animals can make their own vitamin C, but humans cannot. Evolutionary theory says that we share common ancestors with those animals that can make vitamin C, including sharing common ancestors with non-anthropoid primates who can make their own vitamin C.

Evolutionary theory further predicts that since so many different animal species can make vitamin C, it is a distant ancestral state. This means that humans should possess a degraded, nonfunctional version of that ability, and this is what biologists have found. In fact, molecular biology now tells us that humans have a nearly complete biochemical pathway for making vitamin C, except for a mutation in the last step.

Likewise, the structures of hemoglobin and myoglobin show copious evidence of common descent, as do similarities in DNA sequences among animals and people.

The power of evolutionary theory is that with it, all this information fits together, and all these disparate findings make sense.

Powerful Predictions

The power derived from understanding evolution is breathtaking. In the twentieth and twenty-first centuries, evolutionary biology has made predictions that are crucial to human health and well-being. They have come true, sometimes to our peril.

For instance, antibiotic-resistant bacteria are fully predicted by evolutionary theory, and they now threaten the world. Had Darwin himself known about antibiotics, he would have easily predicted that bacteria could evolve defenses against them. In fact, antibiotic-resistant bacteria are now killing people we used to be able to cure.

Meanwhile, the use of genetically modified (GM) agricultural crops has proven to be a laboratory for evolution by natural selection. For instance, glyphosates are a group of powerful herbicides. They are good at killing weeds, but can kill crop plants as well. Genetically modified versions of food crop plants have been created to be resistant to glyphosates. These have been very successful, since farmers can spray large doses of glyphosates on their fields, killing all the weeds while the crops thrive. However, as was inevitable under these circumstances, we are now seeing weed plant populations that have evolved resistance to these powerful herbicides, as was predicted by Darwin's idea of descent with modification.

In both these cases a powerful, beneficial tool has been abused. Antibiotics save human lives and appropriate use of glyphosates can save crop plants that feed humans. But overuse of these tools in defiance of the predictions of evolution by natural selection has put us in a position in which these once-powerful tools become less and less useful over time. We are now in an era in which some disease-causing bacteria have evolved to be resistant to all known antibiotics. Likewise, weed plants are developing multiple forms of glyphosate resistance. These are reducing the usefulness of glyphosates to the point where other means of weed management will soon be required.

However, when evolutionary theory is consulted, humans regain the advantage.

This, for example, has been the case with genetically modified cotton. We now have a type of GM cotton that produces its own insecticide. This enables farmers to grow a successful crop without having to spray dangerous insecticides. However, in time, this would virtually guarantee that resistant strains of insects would evolve. The solution supplied by evolutionary theory is to plant non-GM cotton varieties nearby. This allows populations of the insect pest that have *not* evolved resistance to mate with those that have. The progeny of these matings are very unlikely to be resistant to the GM cotton's internal pesticide. This strategy, known as the refuge strategy, has dramatically slowed or even stopped the development of resistance in insect populations.

We can leave for another day a discussion of whether or not genetically modified crops are beneficial overall. The point here is that evolutionary theory is necessary to wise decision-making, in both agriculture and in public health. I fear having public health officials who do not believe in evolution. I fear having agricultural policy officials who do not believe in evolution. What's more, since ecosystems involve co-evolved relationships between organisms, I fear having environmental officials who don't believe in evolution.

Teaching Evolution

Because the predictions of evolution are so important for human health and wellbeing, evolution must be taught to everybody, regardless of their religious persuasion. For decades, this has not been done. Many educational systems tap-dance around providing rigorous education in evolution for fear of offending religious wishful thinkers who ignore reality, even though the rest of us suffer as a result.

To combat this problem, everyone must be thoroughly educated about evolution and its implications in everyday life. Religious wishful thinking instead of science education must become a thing of the past. When lawmakers and lawyers understand evolution—or at least believe the word of people who do—then we can make scientifically valid public policies, on the many fronts in which evolution by natural selection makes important predictions. When we do this, lives will be saved.

Human Origins

Something must be said on the subject of the "special creation" of human beings. Humans are in general so allergic to seeing themselves as being related to other animals that although most can agree that a domestic cat has a family resemblance to wild lions and tigers, many cannot acknowledge a resemblance between our fellow primates and us. In the 1970s, the field of anthropology was wracked with attempts to find the one, defining thing that made humans different from all other animals. Tool use was suggested and then discarded when zoologists found other animals using tools. Tool making was suggested and kept only until zoologists discovered other animals making tools. Tool use in preparation for the future was then suggested and similarly discarded when other animals were found to make tools in preparation for needing them later. Many humans were intellectually disinclined to see humans as simply another biological species.

Similarly, Christianity was unwilling to see humans as simply another member of the animal kingdom. Adam and Eve being specially created by God and being given dominion over all other animals is a central story in the Bible. Even to this day, at the Creation Museum in Kentucky, there are illustrations showing one family tree for all animals, and a completely separate, single-trunked one for humans, both starting at the dawn of time. No common descent there!

So imagine the significance, then, of Darwin's exposing common descent, and applying that concept to human origins!

In *Origin*, Darwin hinted at the idea that humans are an evolved species. Then, rather than tread lightly around such an explosive topic, he elaborated on it at length in the *Descent of Man*. Further evidence of his bravery and intellectual integrity comes from the fact that in both *Origin* and *Descent* he delved into sexual selection and put forward that idea that natural selection is not entirely based on rudimentary competition and dominance. He introduced the idea that female choice may have something to do with who reproduces and who doesn't. Imagine that! *Females choosing mates* may be just as important as direct male-male physical competition! In fact, the full title of that book is *The Descent of Man, and Selection in Relation to Sex*. No wonder religious fundamentalists don't like evolution! Not only does Genesis go out the window but male domination does too!

An Unlikely Hero

Charles Darwin was an unlikely hero. His health in adulthood was nearly always poor. He was usually seasick while on the *Beagle*, and he was frequently sick throughout his life after the voyage, even to the point of missing crucial meetings. In his education, he switched from medicine to divinity, and then avoided becoming a parson by sailing on the *Beagle*. Much of the time that he was supposed to have spent studying either medicine or divinity was in fact spent studying plants, animals, and geology. Despite his budding prowess as a naturalist, he had to self-fund his trip on the *Beagle*. But he did that crucial thing that scientists do.

He looked at the world carefully with his own eyes.

He took notes, preserved specimens, saw fossil sea shells on mountain tops, saw a familiar piece of ground *raised* as a result of an earthquake that he himself experienced, talked to pigeon breeders, bred pigeons himself, and used copious careful methods that are simply more refined and more precise ways of perceiving things accurately.

Then he reported what he saw, not what he was expected to see. He also reported what he saw rather than what *he himself* had expected to see. He did it with care, bravery, and integrity. As a result, the world was changed forever, for the better.

We are fortunate that Darwin had his apostles. They lent their expertise, their energy, their work, and their own reputations to the mission of having evolution by natural selection rightly accepted by both the scientific community and the world at large. Since intelligent design and creationism are still promoted in the political realm, we cannot say that their victory was complete, but we can say that it was substantial.

Sadly, we live in an era when the idea of scientific truth itself is being disparaged by some of the most powerful people in the world. Scientists themselves are being harassed and their livelihoods and even their lives are being threatened. Today's tenacious evolutionary biologists, climate scientists, and even national park employees are showing themselves to be some of the unlikely heroes of our time. They will need their apostles too. Their apostles are us.

APPENDIX

Chapter 15 Answers to matching game (in State order):

AK: HB2548; Epperson; McLean; AL: HB391; SB336; HB352/SB240; HB106; HB923; HB300; CA: Assn. Of Christian Schools; CO: HB13-1089; FL: SB2692; GA: Selma; IN: HB1356; Hendren; LA: SB561; HB1286; Edward (w/Sup. Ct.); MD: HB1531; MI: 1926 Anti-Evol. Act; HB4382; SB1361; MO: HB1651; HB2554; MT: HB588; NH: HB268; NM: HB506; OH: HB597; HB679; OK: HB2107; PA: Kitzmiller; SC: SB1386; SD: SB112; TN: Butler Act; Scopes Trial; HB368/SB893; TX: HB1485; WA: SB6058; WI: Grantsburg

BIBLIOGRAPHY

1   "Daedalus, or, Science and the Future," a paper read to the Heretics, Cambridge, on February 4, 1923, by J. B. S. Haldane

2   Hesketh, Ian. *Of Apes and Ancestors: Evolution, Christianity and the Oxford Debate* Toronto, University of Toronto Press, 2009), 9.

3   Slotten, Ross A. *Heretic in Darwin's Court: The Life of Alfred Russel Wallace* (New York: Columbia University Press, 2004), 176.

4   "The Great Debate." The Great Debate. Accessed September 15, 2017. www.oum.ox.ac.uk/learning/pdfs/debate.pdf.

5   Hellman, Hal. *Great Fueds in Science: Ten of the Liveliest Disputes Ever* (New York: John Wiley and Sons, 1998), 81.

6   Ibid.

7   Hesketh, Ian. *Of Apes and Ancestors: Evolution, Christianity and the Oxford Debate* Toronto, University of Toronto Press, 2009), 90.

8   Slotten, Ross A. *Heretic in Darwin's Court: The Life of Alfred Russel Wallace* (New York: Columbia University Press, 2004), 175.

9   Ungureanu, James C. *A Yankee at Oxford: John William Draper and the British Association of Science at Oxford, 30 June 1860.* (London: Notes of the Royal Sciety of London, 2016), 135-150.

10 Ibid.

11 Branch, Glenn. "Duane T. Gish Dies." Nation Center for Science Education, March 6, 2013, Accessed November 28, 2017. https://ncse.com/news/2013/03/duane-t-gish-dies-0014753.

12 Dahlgren, Ulric. "The Origin of the Electricity Tissues in Fishes." *The American Naturalist* XLIV, no. 520 (April 1910): 193-202. Accessed July 27, 2018. https://www.jstor.org/stable/2455563?seq=1#metadata_info_tab_contents.

13 Guy, Josephine M. *Victorian Age: An Anthology of Sources and Documents* (London: Routledge, 1998), 273.

14 Gould Stephen Jay. *Bully for Brotosaurus* (New York: Nortion, 1991), 395.

15 Cosslette, Tess. *Science and Religion in the Nineteenth Century.* (Cambridge: Cambridge University Press, 1984), 152.

16 Clark, Ronald W. *The Survival of Charles Darwin-A Biography of a Man and an Idea* (New York: Random House, 1984), 143.

17 Ibid, 143.

18 Clark, Ronald W. *The Survival of Charles Darwin-A Biography of a Man and an Idea* (New York: Random House, 1984), 143-144.

19 Fawcett, Henry. "A Popular Exposition of Mr. Darwin on Origin of Species." *MacMillans*, December 3, 1860, 81-92.

20 "Letter to T.H. Huxley." Charles Robert Darwin to Thomas Henry Huxley. July 20, 1860. In Darwin Correspondence Project. Accessed October 18, 2017. https://www.darwinproject.ac.uk/letter/?docId=letters/DCP-LETT-2873.xml;query=2873;brand=default.

21 "RECORD: Darwin, C.R. and A.R. Wallace." Darwin Online. Accessed November 28, 2017. http://darwin-online.org.uk/content/frameset?itemID=F350&viewtype=text&pageseq=1.

22 Alfred, Randy. "July 1, 1858: Darwin and Wallace Shift the Paradigm." Wired. July 1, 2011. Accessed September 26, 2017. https://www.wired.com/2011/07/0701darwin-wallace-linnaean-society-london/.

23 Frangsmyr, Tore and Lindroth, Sten. *Linnaeus, the Man and His Work* (Los Angeles: University of California Press, 1983), 170.

24 Ibid,169.

25 Repcheck, Jack. *The Man Who Found Time: James Hutton and the Discovery of the Earths Antiquity* (London: Perseus Books, 2003), XX.

26 Ibid, XX.

27 Koestler-Grack, Rachel. *Leonardo da Vinci: Artist, Inventor and Renaissance Man* (Philadelphia: Check House Publishers, 2005), 92.

28 Prothero, Donald R. *Bringing Fossils to Life: An Introduction to Paleobiology* (New York: Columbia University Press, 2013), 51.

29 Rudwick, Martin S.J. *Georges Cuvier, and Geological Cathasthrophism.* (Chicago: University of Chicago Press, 1997), 18.

30 Martin, Ronald. *Earth's Evolving Sytems: The History of Planet Earth* (Boston: James and Bartlett Learning, 2013), 145.

31 "Thomas Malthus (1766–1834)." Introduction to the Aquifoliaceae. Accessed November 29, 2017. http://www.ucmp.berkeley.edu/history/malthus.html.

32 Darwin, Charles. *The Origin of Species By Means of Natural Selection or the Presevaton of Favored Races in the Struggle for Life* (New York: A.L. Burt, 1872), 97.

33 Richards, Robert J. *The Meaning of Evolution: The Morphological Construction and Ideological Reconstruction of Darwin's* Theory (Chicago: University of Chicago Press, 1991), 64.

34 Livingston, David N. and Withers, Charles W.J. *Geographies of Nineteenth-Century Science* (Chicago: University of Chicago Press, 2011), 330.

35 Smith, C.U.M. and Arnott, Robert. *The Genius of Erasmus Darwin* (Hampshire: Ashgate Publishing, 2005), 21.

36 Clark, Ronald W. *The Survival of Chares Darwin-A Biography of a Man and an Idea* (New York: Random House, 1984), 85.

37 Matthew, Patrick. *On Naval Timber and Arborculture, with Critical Notes on Authors Who Have Recently Treated the Subject of Planting* (Edinburgh: Neill & Co., 1831), 364.

38 Clark, Ronald W. *The Survival of Chares Darwin-A Biography of a Man and an Idea* (New York: Random House, 1984), 131.

39 "Letter to Asa Gray." Charles Robert Darwin to Asa Gray. May 22, 1860. In Darwin Correspondence Project. Accessed October 14, 2017. https://www.darwinproject.ac.uk/letter/?docId=letters/ DCP-LETT-2814.xml;query=I own that I cannot see;brand=default.

40 Bird, Alexander. "Thomas Kuhn." Stanford Encyclopedia of Philosophy. October 31, 2018. Accessed November 3, 2018. https://plato.stanford.edu/entries/thomas-kuhn/.

41 Weiner, E.S.C and Simpson, J.A. *Oxford English Dictionary.* Vol. 1. (Oxford: Oxford University Press, 2004).

42 Liu, Joseph. "Public's Views on Human Evolution." Pew Research Center's Religion & Public Life Project. March 14, 2014. Accessed November 24, 2017. http://www.pewforum.org/2013/12/30/publics-views-on-human-evolution/.

43   Tso, Pheoenix. "14 States Use Tax Dollars to Teach Creationism in Public Schools." Mic. October 25, 2015, Accessed December 17, 2019. http://mic.com/articles/80179/14-states-use-tax-dollars-to-teach-creationism-in-public-schools#.quSBS1AV3.

44   Cochrane, Emily. "Years in the Making, Bible Museum Opens in Washington." The New York Times. November 18, 2017. Accessed November 24, 2017. https://www.nytimes.com/2017/11/17/us/politics/bible-museum-hobby-lobby-washington.html.

45   Garson. "Darwinism: Let Us Hope It Is Not True, But If It Is, Let Us Pray It Does Not Become Widely Known." Quote Investigator. Accessed November 24, 2018. https://quoteinvestigator.com/2011/02/09/darwinism-hope-pray/.

46   Dobzhansky, Theodosius. "Nothing in Biology Makes Sense except in the Light of Evolution." *The American Biology Teacher* 35, no. 3 (1973): 125-29. doi:10.2307/4444260.

47   Sargent, John F. *The US Science and Engineering Workforce: Recent, Current and Projected Employment, Wages and Unemployment* (Washington, DC: Congressional Research Service, 2017), 6.

48   Capra, Fritjof. *The Web of Life: A New Scientific Understanding of Living Systems* (New York: Anchor Books/Knopf, 1997). 298.

49   Beck, Alan M. "The Human-Dog Relationship: A Tale of Two Species" in *Dogs, Zoonoses and Public Health.* (Canada: CAB International, 2013), 1-2.

50   Morey, Darcy F. "The Domestic Dog: Its Evolution, Behaviour, and Interactions with People.James Serpell." The Quarterly Review of Biology 72, no. 1 (1997): 87-88.

51   Pollack, Andrew. "Genetically Engineered Salmon Approved for Consumption." The New York Times. January 19, 2018. Accessed January 24, 2018. https://www.nytimes.com/2015/11/20/business/genetically-engineered-salmon-approved-for-consumption.html.

52   Sachs, David H., and Cesare Galli. "Genetic Manipulation in Pigs." *Current Opinion in Organ Transplantation* 14, no. 2 (2009): 148-53. doi:10.1097/mot.0b013e3283292549.

53   Pollack, Andrew. "F.D.A. Approves Drug From Gene-Altered Goats." The New York Times. February 06, 2009. Accessed November 24, 2017. https://www.nytimes.com/2009/02/07/business/07goatdrug.html?

54   Darwin, Charles R. *The Descent of Man, and Selection in Relation to Sex.* (London: John Murray, 1871). 398.

55   *Cosmos: A Personal Voyage.* Performed by Carl Sagan. U.S.A.: Films Inc., 1980. DVD.

56   Brown, Nicola. *The Victorian Supernatural.* (Cambridge: Cambridge University Press, 2004), 2-4.

57   Merill, Lynn M. *The Romance of Victorian Natural History.* (New York: Cambridge University Press, 1989), 55-56.

58   Franklyn, Jeffrey J. *Spirit Matters: Occult Beliefs, Alternative Religion and the Crisis of Faith in Victorian Britain.* (New York: Cornell University Press, 2018),185-187.

59   Lehman, Amy. *Victorian Women and the Theatre of Trance* (London: McFarland and Company, 2009). 145.

60   Quigley, Joan. *What Does Joan Say? My Seven Years as White House Astrologer to Nancy and Ronald Reagan.* (New York: Birch Lane Books, 1990), 82.

61 Levere, Trevor H. *Phrenology and the Origins of Victorian Scientific Naturalism.* (London: Routledge, 2004), 42-43.

62 Marsh, P.T. *The Victorian Church in Decline.* (Pittsburgh: University of Pittsburgh Press, 1969) 124-126.

63 Oppenheim, Janet. *The Other World: Spiritualism and Psychical Research in England, 1810–1914.* (New York: Cambridge University Press, 1985), 63.

64 James, Lawrence. *The Rise and Fall of the British Empire.* (New York: St. Martins Press, 1994), 62.

65 Porter, Andrew. *Oxford History of the British Empire.* Vol. 3. (London: Oxford University Press, 1998), pp.8-9.

66 Levine, Phillippa. *The British Empire: Sunrise to Sunset.* (London: Routledge, 2013), 98-100.

67 Bloom, Paul. *Just Babies: The Origins of Good and Evil.* (New York: Random House, 2013), 39-40.

68 Pinker, Steven. *Better Angels of our Nature: Why Violence Has Declined.* (New York: Penguin Books, 2011), 320.

69 World Happiness Report. "World Happiness Report 2018." World Happiness Report. March 14, 2018. Accessed June 14, 2018. http://worldhappiness.report/ed/2018/.

70 Eppley, Daniel. *Deending Royal Supremacy and Discerning God's Will in Tudor England.* (New York: Routledge, 2007), 14-15.

71 Church History. Accessed September 14, 2017. http://anglican.org/church/ChurchHistory.html.

72 Kieckhefer, Richard. *European Witch Trials: Their Foundations in Popular and Learned Culture, 1300–1500.* (London: Routledge, 2011), .

73 Carlson, Marc. "Witches and Witch Trials in England, the Channel Islands, Ireland and Scotland." Witches and Witch Trials. Accessed January 14, 2017. http://www.personal.utulsa.edu/~marc-carlson/witchtrial/eis.html.

74 Holmes, Ronald. *Witchcraft in British History.* (London: Frederick Miller Ltd., 1974), 51-52.

75 Hatcher, Andrea C. *Political and Religious Identities of British Evangelicals.* (Basingstoke, Hampshire: Palgrave Macmillan, 2018), 113.

76 "One of the World's Richest Organisations Asks for the State to Pay for Its Buildings." Humanists UK. January 19, 2018. Accessed September 22, 2018. https://humanism.org.uk/2018/01/19/new-church-of-england-report-recommends-more-state-funding-for-cathedrals/.

77 Stourton, Edward. *John Paul II: Man of History.* (London: Hodder and Stoughton,   2006), 1.

78 Taylor, Michael W. *Men Versus the State: Herbert Spencer and Late Victorian Individualism.* (London: Oxford University Press, 2006), 245.

79 Haskell, Thomas C. and Teichgraeber, Richard F., III. *The Culture of the Market: Historical Essays.*(Cambridge: Cambridge University Press, 1995). 136-137.

80 Higginbotham, Peter. *Life in Victorian Workhouse* (London: Pitkin Publishing, 2013), 28.

81 Higginbotham, Peter. *Workhouse Encyclopedia* (London: The History Press, 2012), 332.

82 The National Archives. "1833 Factory Act." The National Archives. January 09, 2018. Accessed November 26, 2017. http://www.nationalarchives.gov.uk/education/resources/1833-factory-act/.

83 Hindman, Hugh D. *The World of Child Labor: An Historical and Regional Survey.* (London: M.E. Sharpe, 2009), 554.

84 Knight, Frances. *The Church in the Ninetheenth Century.* (New York: I.B. Travis, 2008), 15-17.

85 Taylor, David. *The New Police in Nineteenth Century England: Crime, Conflict and Control.* (Manchester: Manchester University Press, 1997), 174.

86 Knafla, Louis A. *Crime, Punishment and Reform in Europe.* Vol.18. (Westport: Greenwood Publishing Company, 2003), 114.

87 Mayhew, Henry and Binny, John. *The Criminal Prisons of London and Scenes of Prison Life.* (London: Griffen, Bohn, and Company, 1862), 201.

88 Burtinshaw, Kathryn M. and Burt, John R.F. *Lunatics, Imbeciles and Idiots: A History of Insanity in Nineteenth Century Britain and Ireland.* (London: EPI Group, 2017), 62.

89 Taylor, Paul and Corteen, Karen: *A Companion to Criminal Justice, Mental Health and Risk.* (Bristol: Policy Press, 2014) 27.

90 Steinback, Susie L. *Understanding the Victorians: Politics, Culture and Society in Nineteenth-Century Britain.* (London: Routledge, 2017), 291.

91 Whitbread, Nanette. *The Evolution of the Nursery-Infant School: A History of Infant and Nursery Education in Britain, 1800–1970.* (London: Routledge, 1972), 28.

92 Howe, Anthony. *The Letters of Richard Cobden: 1815–1857.* Vol.1. (Oxford: Oxford University Press, 2007), 165.

93 Coogan, Tim P. *The Famine Plot: England's Role in Ireland's Greatest Tragedy.* (New York: Palgrave-MacMillan, 2012), 31.

94 Clark, Ronald W. *The Survival of Charles Darwin: a Biography of a Man and an Idea.*(New York: Random House, 1984), 21.

95 "What Charles Darwin Read on the 'Beagle." FifteenEightyFour | Cambridge University Press. November 10, 2014. Accessed September 24, 2017. http://www.cambridgeblog.org/2014/11/what-charles-darwin-read-on-the-beagle/.

96 Barlow, Nora. *Charles Darwin and the Voyage of the Beagle.* (London: Pilot Press, 1945), 35.

97 Paley, William. 1743–1805. *Natural Theology: or, the Evidences of the Existence and Attributes of the Diety, Collected from the Appearances of Nature.* (London: R. Faulder, 1803) 1-8.

98 Gross, Charles G. *Huxley versus Owen: the Hippocampus Minor and Evolution.* Trends in Neuroscience Journal (TINS). 16, No.12. 493-497.

99 Bowlby, John. *Charles Darwin: A New Life.* (New York: Norton, 1990), 292-296.

100 Clark, Ronald W. *The Survival of Charles Darwin.* (New York: Random Houses, 1984), 120.

101 "Objections to Origin of Species." Adam Sedgwick to Charles R. Darwin. November 24, 1959. Darwin Correspondence Project (Online).

102 Darwin, Chalres R. 1809–1882. *The Origin of Species: A Variorum Text.* (Philadelphia: University of Pensylvania Press, 1959), p.16

103 Review. *The Saturday Review.* 217, no. 8. 24.12.1859. p.775-776.

104 Bibby, Cyril. *T.H. Huxley: Scientist, Humanist and Educator* (London: Watts, 1959), 3.

105 Clodd, Edward, *Thomas Henry Huxley* (New York: Dodd, Mead and Company, 1902). 3-4

106 Bibby, Cyril. *The Essence of T.H. Huxley: Selections from His Writings* (London: MacMillan, 1967), 4.

107 Clodd, Wdward. *Thomas Henry Huxley* (New York: Dodd, Mead and Company, 1902), 2

108 Clark, Ronald W. *The Huxleys* (New York: McGraw-Hill Book Company), 7-8.

109 Ibid, 35.

110 Bibby, Cyril. *The Essence of T.H. Huxley: Selections from His Writings.* (London: MacMillan, 1967), 34.

111 Rupke, Nicholaas A. *Richard Owen: Biology Without Darwin* (Chicago: University of Chicago Press, 2009), 192.

112 *Proceedings of the Royal Physical Society of Edinburgh* Vol. XII (Edinburgh: Royal Physical Society, 1894), 12.

113 Fahnestock, Jeanne. *Rhetorical Figures in Science* (Oxford: Oxford University Press, 1999), 76.

114 Desmond, Adrian. *Archetypes and Ancestors: Paleontology in Victorian London, 1850–1875,* (Chicago: University of Chicago Press, 1982), 77.

115 *The Correspondence of Charles Darwin: 1858–1859.* Vol 7 (Cambridge: Cambridge University Press, 1991), 391.

116 Manwaring, George. *The Westminster Review.* Vol. 73 (London: Savill and Edwards, 1860), 541.

117 Huxley, Leonard. *Life and Letters of Thomas Henry Huxley.* Vol.1 (London: D. Appleton, 1913), 205.

118 McIntyre, Lee. *Dark Ages: The Case for a Science of Human Behavior* (Cambridge: MIT Press, 2006), 89.

119 Cornish, Charles J. *Sir William Henry Flower, K.C.B* (London: MacMillan, 1904), 66.

120 Cornish, Charles J. *Sir William Henry Flower: A Personal Memoir.* London: Macmillan, 1904.

121 Lyell, Chalres. *The Geological Evidences of the Antiquity of Man, with Remarks on Theories of the Origin of Species* (London: Murray, 1863).

122 Owen, Richard. *On the Anatomy of Vertebrates.* London: Longman's Green, 1866.

123 Straley, Jessica. *Evolution and Imagination in Victorian Children's Literature* (Cambridge: Cambridge University Press, 2016), 81.

124 Clark, Ronald W. *The Huxleys.* (New York: McGraw-Hill, 1969), 65.

125 Bibby, Cyril. *T.H. Huxley: Scientist, Humanist and Educator* (London: Watts, 1959), 248.

126 Ashforth, Albert. *Thomas Henry Huxley* (New York: Twayne Publishers, 1969), 132-133.

127 Wallaston, A.F.R. *The Life of Alfred Newton* (New York: E.P. Dutton and Company, 1921), 102.

128 Linder, Doug. "Thomas Huxley." The Trial of Galileo: An Account. Accessed May 10, 2018. http://law2.umkc.edu/faculty/projects/ftrials/conlaw/huxleyt.html.

129 Ibid.

130 Huxley, Leonard. *Life and Letters of Thomas Henry Huxley.* Vol.1 (New York: D. Appleton, 1913), 420-421.

131 Allen, Mea. *The Hookers of Kew, 1785–1911* (London: Michael Joseph, 1967), 94.

132 Huxley, Leonard. *The Life and Letters of Joseph Dalton Hooker* Vol.1 (Cambridge: Cambridge University Press, 1911), 369.

133 "Letter to Sir. William Jackson Hooker." Joseph Dalton Hooker to William Jackson Hooker. March 2, 1840. In Joseph Dalton Hooker Collection. Accessed March 4, 2018. http://jdhooker.kew.org/p/jdh/asset/1634. JHC

134 "To Joseph Dalton Hooker." Charles Robert Darwin to Joseph Dalton Hooker. January 11, 1844. In Darwin Correspondence Project. Accessed March 20, 2018. https://www.darwinproject.ac.uk/letter/DCP-LETT-729.xml.

135 Prain, David. *Sir Joseph Dalton Hooker, 1817–1911* (Washington: Smithonian Institute, 1912), 660.

136 Endersby, Jim. *Imperial Nature: Joseph Hooker and the Practices of Victorian Science* (Chicago: University of Chicago Press, 2008), 170.

137 El-Hage, Badr "The First Scientific Mission in 1860 to the Cedars of Mount Lebanon" *Archaeology & History in Lebanon* Twelfth Issue: Autumn, 2000, 69-81.

138 Costa, James T. *On the Organic Law of Change* (Cambridge: Harvard University Press, 2013), 431.

139 Clark, Ronald William, and Frederik Pohl. *The Survival of Charles Darwin: A Biography of a Man and an Idea.* (Norwalk: Easton Press, 1991),106.

140 Darwin, Charles R. *The Correspondence of Charles Darwin* Vol.1 (Cambridge: Cambridge University Press, 1991), 166.

141 McCalman, Iain. *Darwins Armada: How Four Voyages to Australasia Won the Battle for Evolution and Changed the World.* (Camberwell, Vic.: Penguin, 2009), 337.

142 "Letter to Charles Darwin." Joseph Dalton Hooker to Charles Robert Darwin. July 2, 1860. In Darwin Correspondence Project. Accessed March 3, 2018.

143 Ibid.

144 Ibid.

145 Ibid.

146 Bean, J.W. *The Royal Botanic Gardens, Kew:Historical and Descriptive* (London: Cassell and Company, 1907), 50.

147 Drayton, Richard. *Nature's Government: Science, Imperial Britain and the 'Improvement' of the World* (New Haven: Yale University Press, 2000), 215.

148 Allen, Mea. *The Hookers of Kew* (London: Michael Joseph, 1967), 208-209.

149 Braislin, William C. *Asa Gray.* Vol. 19 (New York: Medical Society of Brooklyn, 1905), 300.

150 Dana, James D. *Biographical Memoirs of Asa Gray* (Washington: GPO, 1890), p.746.

151 Hoeveller, J. D. *The Evolutionists: American Thinkers Confront Charles Darwin* (New York: Rowman and Littlefield, 2007), 64.

152 Dock, George. *The Medical Library of the University of Michigan* (New York: Medical Library History Journal, 1905), 1.

153 "Letter to Asa Gray." Charles Robert Darwin to Asa Gray. July 20, 1857. In Darwin Correspondence Project. Accessed September 18, 2017. https://www.darwinproject.ac.uk/letter/DCP-LETT-2125.xml.

154 Ibid.

155 "Letter to Charles R. Darwin." Asa Gray to Charles Robert Darwin. August 1857. In Darwin Correspondence Project. Accessed October 14, 2017. https://www.darwinproject.ac.uk/letter/?docId=letters/ DCP-LETT-2129.xml;query=2129;brand=default.

156 Dupree, A. H. *Asa Gray, American Botanist, Friend of Darwin* (Boston, Harvard University Press, 1959), 270-272.

157 Stephenson, Steven C. *The Kingdon Fungi: The Biology of Mushrooms, Molds and Lichens.* (London: Timber Press, 2010), 99.

158 Stamos, David A. *Darwin and the Nature of Species* (New York: SUNY Press, 2007), 135.

159 Dobbs, David. "How Charles Darwin Seduced Asa Gray." Wired. June 03, 2017. Accessed September 24, 2017. https://www.wired.com/2011/04/how-charles-darwin-seduced-asa-gray/.

160 Sargent, Charles S. *The Scientific Papers of Asa Gray,* Vol 2. (Cambridge: Cambridge University Press, 2015), 133.

161 Browne, Janet. *Darwin: The Power of Place* Vol. 2 (New York: Alfred A. Knopf, 2002), 215.

162 "Letter to Charles R. Darwin." Asa Gray to Charles Robert Darwin. July 7, 1865. In Darwin Correspondence Project. Accessed August 17, 2017. https://www.darwinproject.ac.uk/letter/?docId=letters/ DCP-LETT-4877.xml;query=slavery is dead;brand=default.

163 Darwin, Chalres R. *The Descent of Man: And Selection in Relation to Sex* (New York, A.L. Burt Company, 1874), 96-97

164 "Letter to Asa Gray." Charles Robert Darwin to Asa Gray. May 22, 1860. In Darwin Correspondence Project. Accessed October 15, 2017. https://www.darwinproject.ac.uk/letter/?docId=letters/ DCP-LETT-2814.xml;query=appleton gray;brand=default.

165 "Asa Gray." *The Biologist* 52 (1970): 150. Accessed December 14, 2017.

166 https://www.darwinproject.ac.uk/letter/?docId=letters/ DCP-LETT-11330.xml;query=11330;brand=default

167 Asa, Gray. "Natural Selection Not Inconsistent With Natural Theology." Editorial. *Atlantic Monthly*, July 1860, 17. Accessed June 15, 2017. https://www.darwinproject.ac.uk/commentary/religion/essays-reviews-asa-gray/ essay-natural-selection-natural-theology

168 Leahy, Michael P. *Letter to an Atheist* (Tennessee: Harper River Press, 2007), 66.

169 Russett, Cynthia E. *Darwin in America: The Intellectual Response, 1865–1912* (San Francisco: W.H. Freeman and Company, 1976), 8.

170 Hughes, Stefan, *Catches of Light: The Forgotten Lives of the Men and Women Who First Photographed the Heavens* (Cyprus: ArtDeCiel Publishers, 2013), 113.

171 Flemming, Donald. *John William Draper and the Religion of Science* (Philadephia: University of Philadephia Press, 1950), 9-13.

172 *Biographical Memoirs* Vol.2 (Washington: National Academy of Sciences, 1886), 353.

173 Heaton, Claude E. *A Historical Sketch* (New York: New York University College of Medicine, 1841–1941), 9

174 "John William Draper." *Proceedings of the American Academy of Arts and Science* Vol. 85, no. 4 (July 1881): 425-26.

175 Trombino, Donald. "John William Draper." *Journal of the British Anthropological Association* Vol 90, 565-571.

176 Wickcliff, Gregory A. *Draper, Darwin and the Oxford Evolution Debate of 1860. Earth Science and History* Vol. 31, no.1, 1211–1215.

177 Hardin, Jeffrey and Numbers, Ronald. *The Warefare Between Science and Religion: The Idea that Woudn't Die* (Baltimore: Johns Hopkins University Press, 2018), 81.

178 Ungureanu, James C. *A Yankee at Oxford: John William Draper and the British Association of Science at Oxford, 30 June 1860.* (London: Notes of the Royal Sciety of London, 2016), 135-150.

179 "Letter to Charles R. Darwin." Joseph D. Hooker to Charles Robert Darwin. July 2, 1860. In Darwin Correspondence Project. Accessed October 15, 2017. https://www.darwinproject.ac.uk/letter/?docId=letters/ DCP-LETT-2852.xml;query=2852;brand=default.

180 Spencer, Herbert. *The Principles of Biology* (London: Williams and Norgate, 1864). 444.

181 Hesketh, Ian *Of Apes and Ancestors: Evolution, Christianity and the Oxford Debate* (Toronto: University of Toronto Press, 2009), 80.

182 Cosslett, Tess. *Science and Religion in the Nineteenth Century* (Cambridge: Cambridge University Press, 1984), 150.

183 Ungureanu, James C. "A Yankee at Oxford: John William Draper at the British Association for the Advancement of Science at Oxford, 30 June 1860." Collections - Journals | Royal Society. Accessed October 22, 2018. https://royalsocietypublishing.org/doi/full/10.1098/rsnr.2015.0053.

184 Ingersoll, Robert G. *Some Mistakes of Moses* (Washington: C.P.Furrell Publishers, 1880), 22.

185 Holmes, David L. *The Faiths of the Founding Fathers.* (Oxford: Oxford University Press, 2006).

186 Jacoby, Susan. *Freethinkers: A History of American Secularism.* New York: Metro Books, 2004), 13.

187 Foner, Philip S. and Branham, Robert J. *Life Every Voice: African-American Oratory, 1787–1900.* (Tuscaloosa: University of Alabama Press, 1998), 262.

188 Draper, John W. *History of the Conflict Between Religion and Science* (London: Henry S. King and Company, 1875), 217.

189 Foote, George W. "John William Draper." *The Freethinker* (London), January 12, 1896, 16th ed., sec. 1.

190 "Dr. Draper's Lecture on Evolution." *Popular Science Monthly*, December 1877, 175.

191 Raby, Peter. *Alfred Russel Wallace: A Life* (London: Pimlico, 2002), 132.

192 Fichman, Martin. *An Elusive Victorian: The Evolution of Alfred Russel Wallace* (Chicago: University of Chicago Press, 2004), 13.

193 Raby, Peter. *Alfred Russel Wallace: A Life* (London: Pimlico, 2002), 27.

194 Marchant, James. *Alfred Russel Wallace: Letters and Reminiscences* Vol. I (New York: Harper & Brothers, 1916), 94.

195 The Atomism of Democritus. Accessed November 15, 2018. http://people.wku.edu/charles.smith/wallace/BIOG.htm.

196 Smith, Charles H. "The Alfred Russel Wallace Page." The Atomism of Democritus. Accessed November 27, 2017. http://people.wku.edu/charles.smith/wallace/BIOG.htm.

197 Quammen, David. *The Song of the Dodo: Island Biogeography in an Age of Extinctions.* (New York: Scribner, 1996), 18.

198 Wallace, Alfred R. *My Life; A Record of Events and Opinions* Vol.1 (London: Chapman & Hall, 1905), 305.

199 van Whye, John. *Dispelling the Darkness: Voyage in the Malay Archipelago and the Discovery of Evolution by Wallace and Darwin* (New Jersey: World Scientific Press, 2013), 9.

200 Marchant, James. *Alfred Russel Wallace: Letters and Reminiscences* Vol. I (New York: Harper & Brothers, 1916), 78.

201 Fichman, Martin. *An Elusive Victorian: The Evolution of Alfred Russel Wallace* (Chicago: University of Chicago Press, 2004), 74.

202 Wallace, Alfred W. *Natural Selection and Tropical Nature: Essays on Descriptive and Theoretical Biology* (London: MacMillan, 1895), 5.

203 Wallace, Alred W. *Contributions to the Theory of Natural Selection. A Series of Essays* (New York: MacMillan, 1871), 19.

204 Raby, Peter. *Alfred Russel Wallace—A Life* (London: Chatto & Windus, 2001),100-104.

205 "Letter to ARW." Charles Robert Darwin to Alfred Russel Wallace. December 22, 1857. In Darwin Correspondence Project. Accessed September 6, 2017. https://www.darwinproject.ac.uk/letter/?docId=letters/DCP-LETT-2192.xml;query=I am extremely glad;brand=default.

206 Ibid.

207 Young, Robert M. *Darwin's Metaphor: Nature's Place in Victorian Culture* (Cambridge: Cambridge University Press, 1985), 46.

208 Ibid, 46.

209 Darwin, Charles. *The Correspondence of Charles Darwin* Vol.7 (Cambridge: Cambridge University Press, 1991), 108.

210 "Letter to Charles Lyell." Charles Robert Darwin to Charles Lyell. June 18, 1858. In Darwin Correspondence Project. Accessed September 15, 2017. https://www.darwinproject.ac.uk/letter/?docId=letters/DCP-LETT-2285.xml;query=I never saw a more striking;brand=default.

211 Ibid.

212 "Letter to Joseph Hooker." Charles Robert Darwin to Joseph Dalton Hooker. July 7, 1858. In Darwin Correspondence Project. Accessed September 15, 2017. https://www.darwinproject.ac.uk/letter/?docId=letters/DCP-LETT-2303.xml;query=the interest excited was intense;brand=default.

213 Raby, Peter. *Bright Paradise: Victorian Scientific Travellers* (London, Random House, 1996), 169.

214 Wallace, Alred W. *Contributions to the Theory of Natural Selection. A Series of Essays* (New York: MacMillan, 1871), 19.

215 Raby, Peter. *Alfred Russel Wallace: A Life* (London: Pimlico, 2002), 212.

216 "Letter to Charles R. Darwin." Joseph Dalton Hooker to Charles Robert Darwin. October 6, 1865. In Darwin Correspondence Project. Accessed December 11,

        2017. https://www.darwinproject.ac.uk/letter/?docId=letters/DCP-LETT-4910.
        xml;query=Wallace to wonder;brand=default.

217   Smith, Charles H. and Beccaloni, George. *Natural Selection and Beyond: The
        Intellectual Legacy of Alfred Russel Wallace* (Oxford: Oxford University Press,
        2010), 263.

218   Smith, Charles H. "The Alfred Russel Wallace Page." The Atomism of Democri-
        tus. Accessed November 27, 2017. https://people.wku.edu/charles.smith/wal-
        lace/S671.htm.

219   Zinn, Howard. *You Can't be Neutral on a Moving Train.* Boston: Beacon Press.
        2002.

220   *Essays and Reviews.* London: Savill and Edwards. 1860.

221   Wells, Jonathan. *Charles Hodge's Critique of Darwinism: a Historical-Critical
        Analysis of Concepts Basic to the 19th Century Debate.* In Studies in American
        Religion. Vol. 27. (New York: E. Mellen Press, 1988), .

222   Webb, George *The Evolution Controversy in America.* (Kentuky: The University
        Press of Kentucky, 1994), 17.

223   Browne, Janet. *Charles Darwin: The Power of Place.* (New York: Knopf, 2002),
        95.

224   Spurgeon, Charles. "The Gorilla and the Land He Inhabits." Lecture, Metropoli-
        tan Tabernacle, London, October 01, 1861.

225   Artigas, Mariano, Thomas F. Glick, and Rafael A. Martínez. *Negotiating Darwin
        The Vatican Confronts Evolution, 1877–1902.* (Baltimore: Johns Hopkins Univer-
        sity Press, 2009.), 21.

226   Engels, Eve-Marie. *The Reception of Charles Darwin in Europe: Volumes 1 and 2.*
        (London: Continuum, 2014), 416.

227   "Decrees of the First Vatican Council." Papal Encyclicals. December 12, 2017. Ac-
        cessed February 24, 2018. http://www.papalencyclicals.net/councils/ecum20.htm.

228   Angus, Ian. "Marx and Engels...and Darwin?" International Socialist Review.
        Accessed November 24, 2018. https://isreview.org/issue/65/marx-and-engel-
        sand-darwin.

        Issue #65

229   Richarson, John. *Nietzsche's New Darwinism* (London: Oxford University Press,
        2004), 16-18.

230   Wilson, Keith. *A Companion to Thomas Hardy.* (Oxford: Wiley-Blackwell, 2009),
        36.

231   Morgan, Lewis H. *Ancienty Society.* (Chicago: Charles H. Kerr & Company,
        1910), 29-43.

232   Peden, W. C. *From Evolution to Humanism in 19th and 20th Century America.*
        (London: Cambridge Scholars Publishing, 2015), 54.

233   Duke of Argyll. *Darwinism as a Philosophy* in Good Words for 1888. (London:
        Isbister and Company, 1888), 330.

234   Bates, Marston and Humphry, Philip S. *The Darwin Reader.* (New York: Charles
        Scribner's Sons, 1956), 241

235   "Post-Darwin, James Clerk Maxwell's Views Presaged Modern Arguments for
        Intelligent Design." Evolution News. February 08, 2017. Accessed September 20,
        2017. https://evolutionnews.org/2016/12/post-darwin_jam/.

236 "To William Benjamin Carpenter." Charles Robert Darwin to William Benjamin Carpenter. January 6, 1860. In *Darwin Correspondence Project (Online).*Accessed May 22, 2017. https://www.darwinproject.ac.uk/letter/DCP-LETT-2641.xml.

237 Meyer Axel. *Charles Darwin's Reception in Germany and What Followed.* 2009. PLoS Biol 7(7): e1000162. https://doi.org/10.1371/journal.pbio.1000162

238 Browne, Janet. *Charles Darwin: The Power of Place.* (New York: Knopf, 2002), 329.

239 Rectenwald, Michael. *Secularism and the Culture of Nineteenth-Century Scientific Naturalism* in British Journal of the History of Science. 2012. p.16.

240 Tyndall, John. "Address Delivered Before the British Association." Speech, British Association, Belfast, 1874. Accessed February 21, 2017. https://archive.org/details/addressdelivered00tyndrich/page/n5. p.41

241 Jones, Steve. *The Darwin Archipelago.* (New Haven: Yale University Press. 2011), 1

242 Desmond, Adrian and Moore, James. *Darwin: The Life of a Tortured Evolutionist.* (NewYork: Norton, 1992), 488.

243 Stanley, Tim. *The Victorian Era was an Age of Faith – Which is Why it was also a Golden Period of Progress* In. History Today. 61, No. 5. p.4

244 "Church of England Apologizes to Darwin." UPI. September 14, 2008. Accessed November 25, 2017. https://www.upi.com/Church-of-England-apologizes-to-Darwin/52081221412785/.

245 Johnson, Curtis N. *Charles Darwin, Richard Owen and Natural Selection: A Question of Priority.* Jounral of the History of Biology. (2018). https://doi.org/10.1007/s10739-018-9514-2

246 Malthus, Thomas R. *An Essay on the Principles of Population* (New York: Dover Publications), 7-8.

247 Huxley, Thomas Henry. "The Coming of Age of the *Origin of Species*." 1880 Lecture, Royal Institute, London, In *Science.* 2nd ed. Vol. 1. 15-20.

248 Huxley, Thomas H. 1825–1895. *Collected Essays: Darwiniana.* Vol. 2. (New York: D. Appleton and Company, 1893), 247.

249 Bartusiak, Marcia. *Archives of the Universe: A Treasury of Astronomys Historic Works of Discovery.* (New York: Pantheon Books, 2004), 321-322

250 White, Andrew D. 1832–1918. *History of the Warfare of Science with Theology.* Vol. 1. (New York: D. Appleton and Company, 1899), 25.

251 Darwin, Charles R. 1809–1882. *Private Notebook.* Entry 36, 1838.

252 Jacoby, Susan. *Freethinkers: A History of American Secularism.* (New York: Metropolitan/Owl, 2005), 145.

253 Tribe, David H. *100 Years of Freethought.* (London: Elek, 1967), 224.

254 Robertson, J. M. 1856–1933. *A Short History of Freethought.* (London: Watts, 1906), 388

255 Royle, Edward. *Victorian Infidels: The Origins of the British Secularist Movement, 1791–1866.* (Manchester: Manchester University Press, 1974), 276-278.

256 Ingersoll, Robert G. 1833–1899. *The Works of Robert G. Ingersoll in Twelve Volumes.* Volume 2. (New York: Ingersoll League, 1929), 356.

257 Ibid, 358.

258 Ingersoll, Robert G. 1833–1899. *The Works of Robert G. Ingersoll in Twelve Volumes.* Volume 2. (New York: Ingersoll League, 1929), 268.

259　Lewis, Joseph. 1889–1968. *Ingersoll the Magnificent.* (New York: The Free-thought Press Association, 1957), 403.

260　Darwin, Charles R. 1809–1882. *The Descent of Man.* (New York: P.F. Collier & Son, 1902), 361.

261　Stanton, Elizabeth C. 1815–1902. "The Pleasures of Age." 1885 Speech. Suffrage Gathering, New York, NY. July 12, 2017.

262　Stanton, Elizabeth C. 1815–1902. "Bible in Schools" *Evening Post.* (New York), October 10, 1902. Accessed November 12, 1917.

263　Bennett, De Robigne Mortimer. 1818–1882. *A Truth Seeker around the World. A Series of Letters Written While Making a Tour of the Globe.* New York: D. M. Bennett. 1882. Preface.

264　Aveling, E.B. *The Religious Views of Charles Darwin.* (London: Freethought Publishing Company, 1883), .

265ᐧ　White, Andrew D. 1832–1918. *A History of the Warfare of Science with Theology in* Christendom. (New York: D. Appleton and Company, 1896), vii.

266　Ibid. 84.

267　Ibid. 40.

268　"Wants to Subpoena CD." Charles Bradlaugh to Charles R. Darwin. June 5, 1877. Down House, Kent. From Darwin Correspondence Project (online).

269　"Oppose Support of Bradlaugh." Charles Robert Darwin to Charles Bradlaugh. June 6, 1877. Down House, Kent. From Darwin Correspondence Project (online).

270　Cattell, Charles C. "Is Darwin Atheistic." In *Atheistic Platform,* 182. (London: Freethought Publication Company, 1884), .

271　Bradlaugh, Charles. "Some Objections to Socialism." In *Atheistic Platform,* (London: Freethought Publication Company, 1884). 163.

272　Ibid. 168.

273　Aveling, Edward. *The Religious Views of Charles Darwin.* London: Freethought Publication Company, 1883. http://www. darwin-online.org.uk/content/frameset?pageseq=1&itemID=A234&viewtype=side

274　Garrison, H.D. *The Absence of Design in Nature.* (London: Freethought Publication　　　Company, 1884), .

275　Ibid.

276　Besant, Annie. 1847–1933. *The Law of Population: It's Consequences and its Bearing Upon Human Conduct and Morals.* (London: Freethought Publication Company, 1877), 7.

277　Ibid.14

278　Foote, George W. 1850–1915. "PRISONER FOR BLASPHEMY" Accessed November 24, 2016. https://www.guttenberg.org/files/7076=h/7076-h.html.

279　Foote, George W. 1850–1915. *Darwin on God.*(London: Progressive Publishing Company, 1889), 23.

280　Ibid. 32.

281　Ibid. 32.

282　McCabe, Joseph. 1867–1955. *A Biographical Dictionary of Modern Rationalists.* (London: Watts, 1920), 170.

283　Cohen, Chapman. 1868–1954. *An Outline of Evolutionary Ethics.* (London: Forder, 1896), 32.

284 Cohen, Chapman. 1868–1954. *Evolution and Christianity.* (London: Forder, 1897), 9.

285 Darwin, Charles R. 1809–1882. *Journal of Researches into the Natural History and Geology of the Countries Visited During the Voyage of H.M.S. Beagle Round the World.* (New York: D. Appleton and Company, 1871), 500.

286 "Highest to Lowest - Prison Population Rate." Norway | World Prison Brief. Accessed November 20, 2017. http://www.prisonstudies.org/highest-to-lowest/prison_population_rate?field_region_taxonomy_tid=All.

287 "In Depth." BBC News. June 20, 2005. Accessed November 20, 2017. http://news.bbc.co.uk/2/shared/spl/hi/uk/06/prisons/html/nn2page1.stm.

288 Harrington, Rebecca. "These 9 Countries Spend a Greater Share of Money on Science than the United States." Business Insider. March 01, 2016. Accessed November 20, 2017. https://www.businessinsider.com/american-science-funding-statistics-vs-world-2016-2.

289 Amadeo, Kimberly. "How $1 Spent on NASA Adds $10 to the Economy." The Balance Small Business. Accessed December 05, 2018. https://www.thebalance.com/nasa-budget-current-funding-and-history-3306321.

290 Rogers, Simon. "Nasa Budgets: US Spending on Space Travel since 1958 UPDATED." The Guardian. February 01, 2010. Accessed January 22, 2018. https://www.theguardian.com/news/datablog/2010/feb/01/nasa-budgets-us-spending-space-travel.

291 "Historical Trends in Federal R&D." AAAS - The World's Largest General Scientific Society. Accessed August 15, 2018. https://www.aaas.org/programs/r-d-budget-and-policy/historical-trends-federal-rd.

292 Thaxon, Charles and Bradley, Walter L. *The Mysteries of Life's Origin: Reassessing Current Theories.* (Texas: Lewis and Stanley, 1984), .

293 Dawkins, Richard. *The God Delusion.* (New York: Mariner Books, 2008), 322-323.

294 BioLogos. "About BioLogos." BioLogos. Accessed November 18, 2017. https://biologos.org/about-us.

295 Leah Ceccarelli. "Manufactured Scientific Controversy: Science, Rhetoric, and Public Debate." *Rhetoric & Public Affairs* 14, no. 2 (2011): 195-228. https://muse.jhu.edu/ (accessed November 10, 2017).

296 Larson, Edward L. *Summer for the Gods: The Scopes Trial and America's Continuing Debate Over Science and Religion.* (New York: Perseus Books Group, 1997), 107.

297 Kazin, Michael. *A Godly Hero: The Life of William Jennings Bryan.* (New York: Anchor Books, 2007), 178.

298 Shermer, Michael Brant. *The Skeptic Encyclopedia of Pseudoscience.* Vol. 1. (Santa Barbara, CA: ABC-CLIO, 2002), 783.

299 Frater, George. *Our Humanist Heritage: A Handbook for Humanists.* (Longwood, FL: Xulon Pr., 2010), 84.

300 *Theory on the Origin of Man,* H4527 (2002) (testimony of Mike Pence). 107th Congress, 2nd Session. 148, no.93. U.S. House of Representatives

301 Klein, Rebecca. "Husband Of Trump's Education Secretary Once Promoted Intelligent Design In Schools." The Huffington Post. February 22, 2017. Accessed January 22, 2019. https://www.huffingtonpost.com/entry/dick-devos-intelligent-design_us_583b693fe4b000af95ee9046.

302 "PDF Copy of Judge Jones Decision re: Kitzmiller v. Dover." Accessed November 18, 2018. http://www.thelizlibrary.org/evolution-decision.pdf

303   Annenberg Insitute for School Reform. *Public Accountability for Charter Schools: Standards and Policy Recommendations for Effective Oversight.* (Providence: Brown University, 2014), 1-16.

304   Kirk, Chris. "A Map of Thousands of Schools That Are Allowed to Teach Creationism With Taxpayer Money." Slate Magazine. January 26, 2014. Accessed January 25, 2018. http://www.slate.com/articles/health_and_science/science/2014/01/creationism_in_public_schools_mapped_where_tax_money_supports_alternatives.html.

305   "NCES Blog." National Center for Education Statistics (NCES) Home Page, a Part of the U.S. Department of Education. Accessed November 18, 2017. https://nces.ed.gov/blogs/nces/post/a-fresh-look-at-homeschooling-in-the-u-s.

306   Stewart, Katherine. *The Good News Club: The Religious Rights Stealth Assault on Americas Children.* (New York: PublicAffairs, an Imprint of Perseus Books, LLC, a Subsidiary of Hachette Book Group, 2017), 221-222.

307   Chang, Kenneth. "Creationism, Without a Young Earth, Emerges in the Islamic World." *New York Times Online,* November 02, 2009. Accessed December 12, 2018. https://www.nytimes.com/2009/11/03/science/03islam.html.

308   Kingsley, Patrick. "Turkey Drops Evolution From Curriculum, Angering Secularists." The New York Times. June 23, 2017. Accessed November 24, 2017. https://www.nytimes.com/2017/06/23/world/europe/turkey-evolution-high-school-curriculum.html.

309   The Interacademy Panel. *IAP Statement on the Teaching of Evolution.* Italy. 2006. p.1

INDEX